An Introduction to Commutative Algebra

From the Viewpoint of Normalization

An Introduction to Commutative Algebra

From the Viewpoint of Normalization

Huishi Li
Jiaying University, China

World Scientific

NEW JERSEY • LONDON • SINGAPORE • BEIJING • SHANGHAI • HONG KONG • TAIPEI • CHENNAI

Published by

World Scientific Publishing Co. Pte. Ltd.
5 Toh Tuck Link, Singapore 596224
USA office: Suite 202, 1060 Main Street, River Edge, NJ 07661
UK office: 57 Shelton Street, Covent Garden, London WC2H 9HE

British Library Cataloguing-in-Publication Data
A catalogue record for this book is available from the British Library.

AN INTRODUCTION TO COMMUTATIVE ALGEBRA
From the Viewpoint of Normalization

Copyright © 2004 by World Scientific Publishing Co. Pte. Ltd.

All rights reserved. This book, or parts thereof, may not be reproduced in any form or by any means, electronic or mechanical, including photocopying, recording or any information storage and retrieval system now known or to be invented, without written permission from the Publisher.

For photocopying of material in this volume, please pay a copying fee through the Copyright Clearance Center, Inc., 222 Rosewood Drive, Danvers, MA 01923, USA. In this case permission to photocopy is not required from the publisher.

ISBN 981-238-951-2

Printed in Singapore by World Scientific Printers (S) Pte Ltd

For Pinpin and Chao

Preface

Why normalization?

Over the years I had been bothered by selecting material for teaching several one-semester courses. When I taught senior undergraduate students *a first course in (algebraic) number theory*, or when I taught first-year graduate students *an introduction to algebraic geometry*, students strongly felt the lack of some preliminaries on commutative algebra; while I taught first-year graduate students *a course in commutative algebra* by quoting some nontrivial examples from number theory and algebraic geometry, often times I found my students having difficulty understanding. The problem is that in a short semester there is not enough time to go through the significant background material needed in my course (if you are an instructor of mathematics, can you ask your students to find and read those material themselves and do they usually follow?).

Based on my lecture notes on (algebraic) number theory, algebraic geometry, and commutative algebra used at Shaanxi Normal University and Bilkent University, I decided to fuse several things into one — the presentation of this book. As a consequence, the text consists of five chapters that are designed for a (one-semester) common course taken by senior undergraduate students or by first-year graduate students in mathematics. The goal is to introduce to the students the concrete source of commutative algebra through the following diagram:

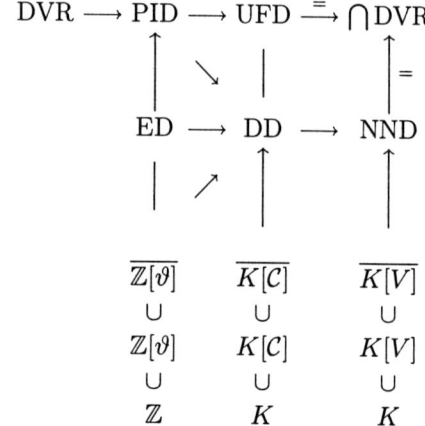

where
- DVR = Discrete valuation ring,
- PID = Principal ideal domain,
- UFD = Unique factorization domain,
- ED = Euclidean domain,
- DD = Dedekind domain,
- NND = Normal Noetherian domain,
- $\overline{\mathbb{Z}[\vartheta]}$ = The integral closure of $\mathbb{Z}[\vartheta]$ (or equivalently the integral closure of $\mathbb{Z}$) in the number field $K = \mathbb{Q}(\vartheta)$,
- $\overline{K[\mathcal{C}]}$ = The normalization (or integral closure) of the coordinate ring $K[\mathcal{C}]$ of an algebraic curve $\mathcal{C}$ in its field of rational functions $K(\mathcal{C})$, and
- $\overline{K[V]}$ = The normalization (or integral closure) of the coordinate ring $K[V]$ of an irreducible algebraic set V in its field of rational functions $K(V)$.

Or more precisely, in terms of *normal (normalized) structure* the lectures demonstrate how to get to the center of commutative algebra by recognizing the roles that $\overline{\mathbb{Z}[\vartheta]}$, $\overline{K[\mathcal{C}]}$ and $\overline{K[V]}$ play in number theory and algebraic geometry, so that, after reading this volume, interested readers may read a course in algebraic number theory at a higher level, or start an advanced course in algebraic geometry with a better background.

I assume that the reader has taken the undergraduate courses including *Linear Algebra* and *A First Course in Abstract Algebra* (Galois theory may not be included). Moreover, very little about topological space is needed to understand the Zariski topology in Chapter 5. Other than these prerequisites, the book is self-contained.

Exercises are given at the end of each section. Though most of the exercises mainly test the understanding of the text in the usual way, the reader is involved in providing proofs and in working problems that have not been completely solved in the text; and furthermore, students are asked to extend some of the theory that is essential for the subsequent sections.

Huishi Li

Contents

Preface   vii

Chapter 1
Preliminaries   1
  0. Conventional Review   1
  1. Noetherian Rings   6
  2. Factorization of Elements in a Domain   9
  3. Field Extensions   18
  4. Symmetric Polynomials   28
  5. Trace and Norm   33
  6. Free Abelian Groups of Finite Rank   40
  7. Noetherian Modules   45

Chapter 2
Local Rings, DVRs, and Localization   53
  1. Spec$R$, m-Spec$R$, and Radicals   53
  2. Local Rings and DVRs   58
  3. The Ring of Fractions and Localization   67
  4. The Module of Fractions   73

Chapter 3
Integral Extensions and Normalization   79
  1. Integral Extensions   79
  2. Noether Normalization   85
  3. Normal Domains and Normalization   90
  4. Normal Domains and DVRs   94

Chapter 4
The Ring $\mathcal{A}_K$ in $K = \mathbb{Q}(\vartheta)$ 99
   1. $\mathcal{A}_K$ is Normal and Free of $\mathbb{Z}$-Rank $[K:\mathbb{Q}]$ 102
   2. $\mathbb{Q}(\sqrt{d})$ and $\mathbb{Q}(\omega)$ 108
   3. Factorization of Elements in $\mathcal{A}_K$ 112
   4. From $\mathcal{A}_K$ to Dedekind Domains   117

Chapter 5
Algebraic Geometry   127
   1. Finite Field Extension and Nullstellensatz   127
   2. Irreducible V and the Prime $\mathbf{I}(V)$ 134
   3. Point P and the Local Ring $\mathcal{O}_{P,V}$ 146
   4. Nonsingular Points and DVRs   151
   5. Normalization of Algebraic Curves   157
   6. Parametrize a Rational Curve via Normalization   162
   7. Rational Curves and Diophantine Equations   167

References   171
Index   173

Chapter 1
Preliminaries

This introductory chapter concentrates on basic notions and properties concerning Noetherian rings, factorization of elements in a domain, field extensions, symmetric polynomials, trace and norm, free abelian groups of finite rank, and Noetherian modules, which might not be familiar to some readers. So we include most of necessary proofs for the reader's convenience.

0. Conventional Review

In this book all rings are *commutative associative* rings with identity 1, and throughout the text,
$\mathbb{N}$ = the set of nonnegative integers,
$\mathbb{Z}$ = the set of integers (ring of integers),
$\mathbb{Z}^+$ = the set of positive integers,
$\mathbb{Q}$ = the set of rational numbers (field of rational numbers),
$\mathbb{R}$ = the set of real numbers (field of real numbers),
$\mathbb{C}$ = the set of complex numbers (field of complex numbers).

Let R be a ring and A a subring of R. Then we insist that A has identity 1_A and

$$1_R = 1_A.$$

And we write

$$R^\times = R - \{0\}.$$

If $\{U_i\}_{i \in \Lambda}$ and $\{V_1, ..., V_m\}$ are collections of nonempty subsets of R, then the sum of $\{U_i\}_{i\in\Lambda}$ and the product of $\{V_1, ..., V_m\}$ are defined as

$$\sum_{i\in\Lambda} U_i = \left\{\sum u_{i_j} \;\middle|\; u_{i_j} \in U_{i_j}\right\},$$

$$V_1 \cdots V_m = \left\{\sum v_1 \cdots v_m \;\middle|\; v_i \in V_i\right\},$$

where the sums involved in both $\sum U_i$ and $V_1 \cdots V_m$ are finite sums. So one understands that
- the sum of given ideals is an ideal;
- the product of finitely many given subrings (ideals) is a subring (an ideal); and
- for subrings (ideals) I, J and K, $I(J+K) = IJ + IK = JI + KI = (J+K)I$.

Let S be a nonempty subset of R and A a subring of R. We set

$\mathbb{Z}[S]$ = the subring of R generated by S
$= \left\{\sum s_{i_1}^{\alpha_1} \cdots s_{i_m}^{\alpha_m} \;\middle|\; s_{i_j} \in S,\; m \in \mathbb{Z}^+,\; \alpha_j \in \mathbb{N}\right\}$

$A[S]$ = the subring of R generated by S over A
$= \left\{\sum a_{(\alpha,j)} s_{i_1}^{\alpha_1} \cdots s_{i_m}^{\alpha_m} \;\middle|\; a_{(\alpha,j)} \in A,\; s_{i_j} \in S,\; m \in \mathbb{Z}^+,\; \alpha_j \in \mathbb{N}\right\}$

$\langle S \rangle$ = the ideal of R generated by S

$= \sum_{s\in S} Rs$

$= \left\{\sum r_i s_i \;\middle|\; r_i \in R,\; s_i \in S\right\},$

where the sums involved in $\mathbb{Z}[S]$, $A[S]$ and $\langle S \rangle$ are finite sums. If $S = \{s_1, ..., s_n\}$ is finite, we write $\mathbb{Z}[S] = \mathbb{Z}[s_1, ..., s_n]$, $A[S] = A[s_1, ..., s_n]$, $\langle S \rangle = \langle s_1, ..., s_n \rangle$, and call them a *finitely generated subring*, a *finitely generated subring over A*, and a *finitely generated ideal* of R, respectively. Clearly we can also write $\langle s_1, ..., s_n \rangle = \sum_{i=1}^n Rs_i$.

Let $R \xrightarrow{\varphi} R'$ be a ring homomorphism. Then we insist that

φ is not the zero-homomorphism and $\varphi(1_R) = 1_{R'}$,

and we write $\text{Ker}\varphi$, $\text{Im}\varphi$ for the kernel and image of φ, respectively. A very useful consequence of the first isomorphism theorem on ring homomorphism states that

- if $R \xrightarrow{\varphi} A$ and $R \xrightarrow{\psi} B$ are ring homomorphisms, φ is surjective, and $\mathrm{Ker}\varphi \subseteq \mathrm{Ker}\psi$, then there is a ring homomorphism $A \xrightarrow{\rho} B$ defined by $\rho(a) = \psi(r)$, where $\varphi(r) = a$, such that the following diagram commutes:

$$\begin{array}{ccc} \mathrm{Ker}\varphi \hookrightarrow R & \xrightarrow{\varphi} & A \\ \psi \downarrow \swarrow \rho & & \rho \circ \varphi = \psi \\ B & & \end{array}$$

if furthermore ψ is surjective, then ρ is surjective as well.

Let R be a ring with identity $1 = 1_R$. If R has no divisors of zero, i.e., $a, b \in R$ and $ab = 0$ implies $a = 0$ or $b = 0$, then R is called an *integral domain*, or simply a *domain*. If R is a domain, then so is the polynomial ring $R[x_1, ..., x_n]$ in variables $x_1, ..., x_n$ over R.

If $a, b \in R$ and $ab = 1$, then a (hence b) is called a *unit* of R. If every nonzero $a \in R$ is a unit, then R is called a *field*.

0.1. Proposition Every finite domain is a field.

Proof Exercise. $\square$

Thus, if $p \in \mathbb{Z}$ is a prime number, then the ring $\mathbb{Z}/\langle p \rangle$ of integers modulo p, usually denoted $\mathbb{Z}_p$, is a field.

Let K be a field. Consider the set of integers

$$o(K) = \left\{ m \in \mathbb{Z}^+ \mid m\lambda = 0 \text{ for some } \lambda \in K^\times \right\}.$$

If $o(K) = \emptyset$, then the *characteristic* of K, denoted $\mathrm{char} K$, is defined to be zero, i.e., $\mathrm{char} K = 0$; if $o(K) \neq \emptyset$, then $\mathrm{char} K$ is defined to be the smallest integer $p \in o(K)$. In the second case p is a prime number (exercise 3).

Every field has a smallest subfield $\mathbb{P}$ (with respect to the inclusion relation on subfields), the *prime subfield*, which is either isomorphic to

$$\mathbb{Q}, \text{ if } \mathrm{char} K = 0,$$

or to

$$\mathbb{Z}_p, \text{ if } \mathrm{char} K = p > 0.$$

Clearly, every finite field F has $\mathrm{char} F > 0$. If a field K has $\mathrm{char} K = p > 0$, then $(a + b)^p = a^p + b^p$ for all $a, b \in K$ (exercise 4).

If R is a domain, then the *field of fractions* of R is constructed via the equivalence relation on $R \times R^\times$:

$(a, b) \sim (c, d)$ if and only if $bc = ad$.

Write $\frac{a}{b}$ for the equivalence class represented by (a, b), and write $Q(R)$ for the quotient set $R \times R^\times / \sim$. Then

$$Q(R) = \left\{ \frac{a}{b} \;\middle|\; a, b \in R,\ b \neq 0 \right\}$$

where the addition and multiplication are defined the same as that for rational numbers. Thus, in $Q(R)$, $\frac{0}{1} = 0$ is the zero of the additive group $(Q(R), +)$, $\frac{1}{1} = 1_{Q(R)}$ is the identity of the multiplicative group $(Q(R), \cdot)$, and if $\alpha = \frac{a}{b} \neq 0$ then $\alpha^{-1} = \frac{b}{a}$.

The ring homomorphism

$$\lambda_R : R \longrightarrow Q(R)$$

$$r \mapsto \frac{r}{1}$$

is injective. In the case where R is a field, λ_R is an isomorphism. So we may view R as a subring of $Q(R)$ and write $R \subseteq Q(R)$. Consequently, if $Q(R)[x]$ is the polynomial ring in variable x over $Q(R)$, then $R[x] \subseteq Q(R)[x]$.

If R' is another domain and $\varphi: R \to R'$ is an injective ring homomorphism, then φ induces an injective ring homomorphism $\overline{\varphi}: Q(R) \to Q(R')$, where $\overline{\varphi}\left(\frac{a}{b}\right) = \frac{\varphi(a)}{\varphi(b)}$. Hence $Q(R)$ may be viewed as a subfield of $Q(R')$, and consequently $Q(R)[x]$ may be viewed as a subring of $Q(R')[x]$. It turns out that if φ is an isomorphism then so is $\overline{\varphi}$. In particular, if $Q(R)$ is the field of fractions of the domain R and $R \subset B \subset Q(R)$, where B is a subring of $Q(R)$, then $Q(B) = Q(R)$.

We assume that the reader is familiar with the basic structural properties of a polynomial ring $R[x_1, ..., x_n]$ in variables $x_1, ..., x_n$ over the ring R, for instance, R is a subring of $R[x_1, ..., x_n]$ consisting of constant polynomials, every $f \in R[x_1, ..., x_n]$ has a *unique* expression into the linear combination of monomials: $f = \sum r_\alpha x_1^{\alpha_1} \cdots x_n^{\alpha_n}$, where $\alpha = (\alpha_1, ..., \alpha_n) \in \mathbb{N}^n = \{(\alpha_1, ..., \alpha_n) \mid \alpha_1, ..., \alpha_n \in \mathbb{N}\}$, and the *degree* of f is defined as

$$\deg f = \max \left\{ \alpha_1 + \cdots + \alpha_n \;\middle|\; r_\alpha x_1^{\alpha_1} \cdots x_n^{\alpha_n} \neq 0 \text{ is a term of } f \right\}.$$

If $f = 0$ then conventionally $\deg f$ is defined as $-\infty$. Thus, for $f = \sum r_\alpha x_1^{\alpha_1} \cdots x_n^{\alpha_n}$, $g = \sum r_\beta x_1^{\beta_1} \cdots x_n^{\beta_n}$, one knows how to determine the degree of $f + g$ and $f \cdot g$ according to the addition and multiplication of polynomials.

In particular, we recall the following important properties of a polynomial ring.

If $f \in R[x]$ is a nonconstant *monic polynomial*, i.e., $\deg f \geq 1$ and f is of the form

$$f = x^n + a_{n-1}x^{n-1} + \cdots + a_0, \quad a_i \in R,$$

then a division algorithm on $g \in R[x]$ by f exists:

$$g = qf + r, \quad q, r \in R[x], \ \deg r < \deg f.$$

Let R be a ring. If $\mathbb{Z}[s_1, ..., s_n]$ is the subring of R generated by $s_1, ..., s_n$, then there is an onto ring homomorphism from the polynomial ring $\mathbb{Z}[x_1, ..., x_n]$ to $\mathbb{Z}[s_1, ..., s_n]$:

$$\mathbb{Z}[x_1, ..., x_n] \longrightarrow \mathbb{Z}[s_1, ..., s_n]$$

$$f(x_1, ..., x_n) \mapsto f(s_1, ..., s_n)$$

Let A be a subring of R. If $A[s_1, ..., s_n]$ is the subring of R generated by $s_1, ..., s_n$ over A, then there is an onto ring homomorphism from the polynomial ring $A[x_1, ..., x_n]$ to $A[s_1, ..., s_n]$:

$$A[x_1, ..., x_n] \longrightarrow A[s_1, ..., s_n]$$

$$f(x_1, ..., x_n) \mapsto f(s_1, ..., s_n)$$

Exercises

1. Let A be a subring of the ring R and $S \subseteq R$ a nonempty subset of R. Show that $\mathbb{Z}[S]$ is the smallest subring of R containing S, that $A[S]$ is the smallest subring of R containing A and S, and that $\langle S \rangle$ is the smallest ideal of R containing S. (Here the ordering on subrings and ideals is the usual inclusion ordering on subsets.)
2. Complete the proof of Proposition 0.1. (Hint: If $R = \{a_1, ..., a_n\}$ is a finite domain and $0 \neq a_i \in R$, then $a_i R = R$.)
3. Let K be a field. Show that if $\operatorname{char} K = p \neq 0$, then p is a prime number.
4. Let K be a field of $\operatorname{char} K = p \neq 0$. Show that $(a + b)^p = a^p + b^p$ for all $a, b \in K$. (Hint: Check the binomial coefficients of the expansion of $(a + b)^p$.)

1. Noetherian Rings

Since Noetherian ring plays a leading role in commutative algebra, we start with this notion.

Let R be a ring. R is said to satisfy the *maximal condition* if every nonempty set of ideals contains a maximal member with respect to the inclusion relation on ideals. R is said to satisfy the *ascending chain condition* if for every ascending chain of ideals

$$I_1 \subseteq I_2 \subseteq \cdots \subseteq I_n \subseteq \cdots$$

there is some k such that $I_k = I_j$ for all $j \geq k$.

1.1. Theorem Let R be a ring. The following are equivalent.
(i) R satisfies the maximal condition.
(ii) Every ideal of R is finitely generated.
(iii) R satisfies the ascending chain condition.

Proof (i) $\Rightarrow$ (ii) Let I be a nonzero ideal of R. Set

$$S = \{\text{all finitely generated ideals contained in } I\}.$$

Then $S \neq \emptyset$, and by (i) there is a maximal member in S, say $J = \sum_{i=1}^{n} Ra_i$ with $a_i \in I$. If $J \neq I$, then there is some $x \in I$, $x \notin J$. Thus, J is properly contained in $J' = J + Rx$ and $J' \in S$, contradicting the choice of J. Therefore $I = J$, a finitely generated ideal.

(ii) $\Rightarrow$ (iii) Let

$$I_1 \subseteq I_2 \subseteq \cdots I_n \subseteq \cdots$$

be an ascending chain of ideals in R. Set $I = \cup I_i$. Then I is an ideal of R and hence finitely generated, say $I = \sum_{j=1}^{m} Ra_j$ with $a_j \in I$. Suppose $a_j \in I_{i_j}$ with $i_1 < i_2 < \cdots < i_m$. Then $a_j \in I_{i_m}$, $j = 1, ..., m$, and consequently $I = I_{i_m}$. Let $k = i_m$. Then $I_k = I_j$ for all $j \geq k$.

(iii) $\Rightarrow$ (i) Let $S = \{I_i\}$ be a nonempty set of ideals in R. If S did not have a maximal member, there would be a strictly ascending chain of ideals out of S, which does not satisfy the chain condition. □

1.2. Definition A ring R satisfying one of the equivalent conditions of Theorem 1.1 is called a *Noetherian ring*.

Let R be a ring. If every ideal I of R is a principal ideal, i.e., $I = \langle a \rangle = Ra$ for some $a \in I$, then R is called a *principal ideal ring*. Principal ideal rings are special Noetherian rings. If a principal ideal ring R is also a domain, then we simply call R a PID.

It is a result of the division algorithm in $\mathbb{Z}$ and the division algorithm in the polynomial ring $K[x]$, where K is a field, that both $\mathbb{Z}$ and $k[x]$ are PIDs (exercise 1).

Concerning polynomial rings in finitely many variables over a Noetherian ring, we have the following celebrated result.

1.3. Theorem (Hilbert basis theorem) If R is a Noetherian ring then so is the polynomial ring $R[x]$ in variable x over R. Hence, the polynomial ring $R[x_1, ..., x_n]$, in any finitely n variables $x_1, ..., x_n$, is Noetherian.

Proof We show that if $R[x]$ is not Noetherian then neither is R, by adopting a well-known argumentation (as one may easily find at the site [1]).

Suppose that I is an ideal of $R[x]$ which is not finitely generated. Then a sequence of polynomials from I can be chosen as follows.

$$f_1 \in I \text{ with least degree } n_1,$$
$$f_2 \in I - Rf_1 \text{ with least degree } n_2,$$
$$f_3 \in I - (Rf_1 + Rf_2) \text{ with least degree } n_3,$$
$$\vdots$$
$$f_{k+1} \in I - \sum_{i=1}^{k} Rf_i \text{ with least degree } n_{k+1},$$
$$\vdots$$

where $n_1 \leq n_2 \leq n_3 \leq \cdots \leq n_{k+1} \leq \cdots$.

Claim Let a_i be the leading coefficient of f_i. Then

$$Ra_1 \subset Ra_1 + Ra_2 \subset \cdots \subset \sum_{i=1}^{n} Ra_i \subset \cdots$$

is a strictly ascending chain of ideals in R.

If the claim was not true, then $\sum_{i=1}^{k} Ra_i = \sum_{i=1}^{k+1} Ra_i$ for some k, and this would yield $a_{k+1} = \sum_{i=1}^{k} r_i a_i$, $r_i \in R$. Note that for $i = 1, ..., k$, we

---
[1] http://planetmath.org/encyclopedia/ProofofHilbertBasisTheorem.html

have

$$f_i = a_i x^{n_i} + \text{strictly lower degree terms,}$$
$$r_i f_i x^{n_{k+1}-n_i} = r_i a_i x^{n_{k+1}} + \text{strictly lower degree terms.}$$

It follows that

$$\sum_{i=1}^{k} r_i f_i x^{n_{k+1}-n_i} = \left(\sum_{i=1}^{k} r_i a_i\right) x^{n_{k+1}} + g(x)$$

$$= a_{k+1} x^{n_{k+1}} + g(x),$$

while

$$g(x) = \left(f_{k+1} - \sum_{i=1}^{k} r_i f_i x^{n_{k+1}-n_i}\right) \notin \sum_{i=1}^{k} R f_i$$

by the choice of f_{k+1}. But clearly $\deg g(x) < \deg f_{k+1}$, contradicting the choice of f_{k+1}. Therefore the claim holds, i.e., R is not Noetherian. □

The polynomial ring $K[x_1, ..., x_n, ...]$ in infinitely many variables over a field K is non-Noetherian, due to the existence of a strictly ascending chain of ideals:

$$\langle x_1 \rangle \subset \langle x_1, x_2 \rangle \subset \cdots \subset \langle x_1, ..., x_n \rangle \subset \cdots .$$

In Chapter 4 we will see that if $\mathcal{A}$ is the set of all algebraic integers, i.e., the set of complex zeros of monic polynomials in $\mathbb{Z}[x]$, then $\mathcal{A}$ forms a ring and it is not Noetherian; while for a finite dimensional field extension $\mathbb{Q} \subseteq K$ with K a subfield of $\mathbb{C}$, $\mathcal{A} \cap K$ is always Noetherian.

Noetherian rings stemming from algebraic geometry are given in Chapter 5.

Exercises

1. Show that $\mathbb{Z}$ and $K[x]$ are PIDs, where $K[x]$ is the polynomial ring in x over a field K.
2. Let $R \to R'$ be an onto ring homomorphism. Show that if R is Noetherian then so is R'.
3. Let A be a Noetherian subring of the ring R, and let $\{r_1, ..., r_s\}$ be a finite subset of R. Show that the subring $A[r_1, ..., r_s]$ of R is Noetherian.
4. Let K be a field which, as a subring, is contained in the ring R. Assume that R is finite dimensional over K. Show that R is Noetherian.

5. Let R be a Noetherian ring. The ring of formal power series over R is the associative ring $R[[x]]$ consisting of the formal series

$$f(x) = \sum_{i=0}^{\infty} r_i x^i, \quad r_i \in R,$$

where $f(x) = 0$ if and only if $r_i = 0$ for all $i = 0, 1, \ldots$, and the addition and multiplication are defined as for the power series with real coefficients in calculus. Show that $R[[x]]$ is Noetherian. (Hint: Define the degree of a series as the lowest power of x.)

6. By Theorem 1.3, $\mathbb{Z}[x]$ is Noetherian. Show that the ideal $I = \langle 2, x \rangle$ is not a principal ideal.

7. Let $\mathbb{Z}_2[x, y]$ be the polynomial ring over the field $\mathbb{Z}_2$. Show that in $\mathbb{Z}_2[x, y]/\langle x^2 + x + y^3 + 1 \rangle$ the ideal $\langle \overline{x}, \overline{y+1} \rangle$ is not a principal ideal.

2. Factorization of Elements in a Domain

Let R be a domain. It is easy to see that the set of units in R, denoted

$$U(R) = \left\{ u \in R \mid u \text{ is a unit in } R \right\},$$

forms a group with respect to the multiplication of R. $U(R)$ is called the *group of units* in R.

2.1. Definition (i) For $r \in R$, $u \in U(R)$, the element $y = ur = ru$ is called an *associate* of r.

(ii) Let $r, s \in R$. r is said to be *divisible* by s, denoted $s|r$, if $r = sz$ for some $z \in R$, where s (hence z) is called a *divisor* (or a factor) of r.

For $u \in U(R)$ and $r \in R$, u and ur are called the *trivial divisors* of r (note that $r = (ur)u^{-1} = (u^{-1}r)u$).

(iii) For $r \in R$, if r has only trivial divisors in R, then we say that r is *irreducible* in R; otherwise, r is *reducible* in R. (So zero is reducible in any domain.)

(iv) For $r \in R$, if r is reducible, then $r = sz$ with nontrivial divisors s, z. In this case we say that r has a *proper factorization*.

Example (i) Let $R = \mathbb{Z}$. Then $U(R) = \{\pm 1\}$.

(ii) Let $R = \mathbb{Z}[i]$ where $i = \sqrt{-1}$. Then $U(R) = \{\pm 1, \pm i\}$ (see Chapter 4 section 3).

(iii) Let $R = K[x]$ be the polynomial ring in x over a field K. Then $U(R) = K^\times$.

Thus, one easily finds elements in each R that have proper factorization.

2.2. Proposition Let R be a domain, $r, s \in R$. The following hold:
(i) $r \in R$ is a unit if and only if $r|1$.
(ii) Any two units are associates to each other, and any associate of a unit is a unit.
(iii) r, s are associates to each other if and only if $r|s$ and $s|r$.
(iv) r is irreducible if and only if every divisor of r is either an associate of r or a unit.
(v) Any associate of an irreducible element is irreducible.

Proof Exercise. □

In terms of ideal structure, we may characterize units, divisibility, associates and irreducibility, as follows.

2.3. Proposition Let R be a domain and let r, s be nonzero elements of R.
(i) $r \in U(R)$ if and only if $\langle r \rangle = R$.
(ii) $r|s$ if and only if $\langle r \rangle \supseteq \langle s \rangle$.
(iii) r, s are associates to each other if and only if $\langle r \rangle = \langle s \rangle$.
(iv) r is irreducible if and only if $\langle r \rangle$ is maximal among the principal ideals of R (with respect to the inclusion ordering on ideals).

Proof Exercise. □

2.4. Definition Let R be a domain. We say that *factorization into irreducible elements is feasible* in R if every nonzero nonunit element may be expressed as a product of finitely many irreducible elements.

2.5. Proposition Factorization into irreducible elements is feasible in a Noetherian domain R.

Proof Let R be a Noetherian domain. Suppose that the assertion was not true. Then the set Ω of nonzero nonunit elements which cannot be factorized into finite products of irreducible elements would be nonempty.

Since R is Noetherian, let $\langle y \rangle$ be a maximal member in

$$S = \left\{ \langle x \rangle \mid x \in \Omega \right\}.$$

Then y is reducible because $y \in \Omega$, and $y = rs$ for $r, s \notin U(R)$. Thus, $\langle y \rangle$ is properly contained in $\langle r \rangle \cap \langle s \rangle$ (otherwise r or s would be a unit by Proposition 2.3). By the choice of $\langle y \rangle$ we have

$$r = p_1 \cdots p_r, \qquad s = p_{r+1} \cdots p_n$$

where p_i's are irreducible elements. But then $y = p_1 \cdots p_r p_{r+1} \cdots p_n$, a product of finitely many irreducible elements. This is a contradiction and hence $\Omega = \emptyset$. □

2.6. Definition Let R be a domain in which factorization into irreducible elements is feasible. For a nonzero nonunit $x \in R$, if any two factorizations

$$x = p_1 \cdots p_n \text{ and } x = q_1 \cdots q_m$$

satisfy $n = m$ and (up to the arrangement of divisors) $p_i = u_i q_i$, $i = 1, ..., n$, where $u_i \in U(R)$, then x is said to have a *unique* factorization in R. If every nonzero nonunit element of R has a unique factorization in R, we say that R is a UFD (abbreviation of the phrase "unique factorization domain").

Remark At this stage, it is better to be aware of two facts.
(i) There are Noetherian domains which are not UFDs (see exercise 4 of this section and Chapter 4 section 3).
(ii) There are UFDs which are not Noetherian (exercise 5).

In order to discuss the uniqueness of factorization into irreducible elements, we introduce the notion of a prime in a domain.

2.7. Definition Let R be a domain, $0 \neq x \in R$, and $x \notin U(R)$. x is said to be a *prime* if $x|ab$ implies $x|a$ or $x|b$ for any $a, b \in R$.

2.8. Proposition Let p be a prime in a domain R. The following hold:
(i) Any associate of p is a prime in R.
(ii) p is irreducible in R.

Proof Exercise. □

2.9. Theorem If factorization into irreducible elements is feasible in a domain R, then R is a UFD if and only if every irreducible element is a prime.

Proof Since factorization into irreducible elements is feasible in R, by Proposition 2.2(v), every nonzero nonunit $x \in R$ has a factorization

$$x = p_1 \cdots p_\ell,$$

where p_i may be an associate of some irreducible element.

First suppose that factorization in R is unique. Let p be an irreducible element and $p|ab$ where $a \neq 0$, $b \neq 0$. Then $ab = pc$ for some $0 \neq c \in R$. Consider the unique factorizations: $a = p_1 \cdots p_n$, $b = q_1 \cdots q_m$, $c = r_1 \cdots r_s$. Then

$$pc = p(r_1, ..., r_s) = (p_1 \cdots p_n)(q_1 \cdots q_m) = ab.$$

By the uniqueness, p divides some p_i or some q_j. Hence $p|a$ or $p|b$, and this shows that p is a prime.

Conversely, suppose every irreducible element is a prime. Consider the factorization into primes

$$x = p_1 \cdots p_n = q_1 \cdots q_m.$$

Then $p_1|q_1(q_2 \cdots q_m)$. Without loss of generality we may assume $p_1|q_1$. Then, $q_1 = u_1 p_1$ for some $u_1 \in U(R)$ because q_1 has only trivial divisors. Thus, $x = p_1 \cdots p_n = (u_1 p_1)(q_2 \cdots q_m)$ and $p_2 \cdots p_n = (u_1 q_2)(q_3 \cdots q_m)$. After repeating this process n times, up to the arrangements of divisors we derive $q_i = u_i p_i$ with $u_i \in U(R)$, and $m \leq n$. Similarly, $n \leq m$. So $n = m$. This shows that factorization is unique in R. □

2.10. Theorem Every PID is a UFD.

Proof Let R be a PID. Then factorization into irreducible elements is feasible in R because R is Noetherian.

Let p be an irreducible element in R. Then by Proposition 2.3(iv), $\langle p \rangle$ is maximal among all ideals. Suppose $p|ab$ but $p \nmid a$. Then $\langle p \rangle$ is properly contained in the ideal $\langle p, a \rangle$. By the maximality of $\langle p \rangle$ we have $\langle p, a \rangle = R$. It follows that $1 = ph + aq$ and $b = bph + abq$. This yields $p|b$, showing that p is a prime. By Theorem 2.9, R is a UFD. □

Remark Recall that before learning a systematic theory on UFDs, in the arithmetic theory on $R = \mathbb{Z}$ (or in $R = K[x]$ where K is a field) a prime p is defined as the element which has only the divisors ± 1 ($\lambda \in K^\times$), $\pm p$ (λp). If $a, b \in R$, $p | ab$ but $p \nmid a$, then the Euclidean algorithm output the greatest common divisor $\gcd(p, a) = 1$ in the form

$$af + pg = 1, \quad f, g \in R,$$

that yields $p | b$ immediately as in the above proof. That is why we know, without arguing that R is a PID, that R is a UFD. Indeed, there is a class of UFDs that hold a version of Euclidean algorithm, as described below.

2.11. Definition A Euclidean domain is a domain R with a function (called a Euclidean function):

$$\phi : \quad R^\times \longrightarrow \mathbb{N}$$

that satisfies
(i) if $a, b \in R^\times$ and $a | b$ then $\phi(a) \leq \phi(b)$; and
(ii) if $a, b \in R^\times$ then there exist $q, r \in R$ such that

$$a = qb + r, \text{ where either } r = 0 \text{ or } \phi(r) < \phi(b).$$

Example (iv) $\mathbb{Z}$ is a Euclidean domain with the Euclidean function given by the absolute value function. $K[x]$ is a Euclidean domain with the Euclidean function given by the degree function. (A consequence of applying the division algorithm to both $\mathbb{Z}$ and $K[x]$.)

2.12. Theorem Every Euclidean domain R is a PID.

Proof Let I be a nonzero ideal of R. If ϕ is the associated Euclidean function on R, let us set

$$\phi(x^*) = \min \left\{ \phi(x) \in \mathbb{N} \ \middle| \ 0 \neq x \in I \right\}.$$

For any $0 \neq y \in I$, $y = qx^* + r$ with $r = 0$ or $\phi(r) < \phi(x^*)$. But $r = y - qx^* \in I$. By the choice of x^*, $r = 0$. Thus, $y = qx^*$. This shows that $I = \langle x^* \rangle$. $\square$

2.13. Corollary Every Euclidean domain is a UFD.

Proof This follows from Theorems 2.10–2.12. $\square$

Except for $\mathbb{Z}$ and $K[x]$, other Euclidean domains will be given in Chapter 4 section 3.

Remark Let $K = \mathbb{Q}(\sqrt{-19})$. By Theorem 3.4 of Chapter 4, the ring $\mathcal{A}_K$ of algebraic integers in K is not a Euclidean domain. However, $\mathcal{A}_K$ is a PID. The reader is referred to http://www.mathreference.com/id,npid.html for a beautiful proof on this fact.

We now proceed to show that the polynomial ring $R[x]$ in variable x over a UFD R is a UFD.

2.14. Lemma (Gauss) Let R be a domain. Then any prime of R is a prime in $R[x]$.

Proof Let p be a prime in R and $\overline{R} = R/\langle p \rangle$. Then a direct verification shows that $\overline{R}$ is a domain, and so is the polynomial ring $\overline{R}[x]$. If $f, g \in R[x]$ and $p|fg$, then $fg \in \langle p \rangle$. For $r \in R$, write $\overline{r}$ for the image of r in $\overline{R}$. Consider the ring homomorphism

$$R[x] \xrightarrow{\varphi} \overline{R}[x]$$

$$\sum r_i x^i \mapsto \sum \overline{r_i} x^i$$

Then $\varphi(fg) = \overline{f} \cdot \overline{g} = 0$. Since $\overline{R}[x]$ is a domain, it follows that $\overline{f} = 0$ or $\overline{g} = 0$, i.e., $p|f$ or $p|g$, as desired. □

Let R be a UFD. Then for any $r_1, ..., r_n \in R$, not all zero, the greatest common divisor $\gcd(a_1, ..., a_n)$ exists in R (exercise 6).

2.15. Definition Let R be a UFD. If a polynomial $r_n x^n + r_{n-1} x^{n-1} + \cdots + r_0 = f(x) \in R[x]$ has the property that $\gcd(r_n, r_{n-1}, ..., r_0) = d \in U(R)$, then $f(x)$ is called a *primitive polynomial*.

2.16. Proposition Let R be a UFD. If $f, g \in R[x]$ are primitive then so is the product fg.

Proof This follows immediately from Gauss lemma. □

2.17. Theorem Let R be a UFD with the field of fractions $K = Q(R)$.
(i) If $f \in R[x]$ and $f = gh$ for some $g, h \in K[x]$, then there is a unit

$\alpha \in K[x]$ such that $g\alpha, \alpha^{-1}h \in R[x]$.

(ii) Let $f, g \in R[x]$, where g is primitive. If $g|f$ in $K[x]$ then $g|f$ in $R[x]$.

Proof (i) Let $f = gh$ be as assumed. Let $r_0 \in R$ be the common denominator of the coefficients of g. Then $r_0 g \in R[x]$. Let d be the greatest common divisor of all coefficients of $r_0 g$. Then $g_1 = \alpha g$ is primitive in $R[x]$, where $\alpha = \frac{r_0}{d} \in K$. Similarly, there exists $\beta \in K$ such that $h_1 = \beta h$ is primitive in $R[x]$. Set $\alpha\beta = \frac{a}{b}$, where a and b have only common divisors which come from $U(R)$. Then

$$\frac{a}{b}f = \alpha\beta gh = g_1 h_1 \text{ and } af = bg_1 h_1.$$

Now, if $a \in U(R)$, then since $b\alpha\beta = a$, we have $\alpha g = g_1$, $\alpha^{-1}h = a^{-1}b\beta h = a^{-1}bh_1$ have coefficients in R, as desired. So it remains to show that $a \in U(R)$. If not, there would be some prime p dividing a. Hence $p|bg_1 h_1$ but $p \nmid b$ by the choice of a and b, and $p \nmid g_1$, $p \nmid h_1$ because both g_1 and h_1 are primitive. This contradicts Gauss lemma. Therefore, a must be a unit.

(ii) This follows from part (i). □

Let R be a UFD and $f(x) \in R[x]$ with $\deg f(x) \geq 1$. If d is the greatest common divisor of all coefficients of $f(x)$, then $f(x) = df_1(x)$ where $f_1(x)$ is a primitive polynomial. Bearing this fact in mind, Theorem 2.17 enables us to derive immediately the following.

2.18. Proposition Let R be a UFD with the field of fractions $K = Q(R)$, $p(x) \in R[x]$ with $\deg p(x) \geq 1$. Then $p(x)$ is irreducible in $R[x]$ if and only if $p(x)$ is irreducible in $K[x]$.

□

2.19. Theorem If R is a UFD then so is $R[x]$.

Proof Since R is a UFD and $R \subset R[x]$, by Gauss lemma we need only to consider polynomials of degree ≥ 1.

Let $K = Q(R)$ be the field of fractions of R. Then $K[x]$ is a UFD. Thus every $f(x) \in R[x]$ with $\deg f(x) \geq 1$ is factorized into a product of finitely many irreducible elements in $K[x]$. By Theorem 2.17 and Proposition 2.18, factorization of polynomials of degree ≥ 1 into irreducible polynomials is feasible in $R[x]$, and irreducible polynomials in $R[x]$ are primes. Hence $R[x]$ is a UFD. □

2.20. Corollary For any field K, the polynomial ring $K[x_1, ..., x_n]$ in finitely many variables $x_1, ..., x_n$ over K is a UFD. □

We finish this section by Eisenstein's criterion concerning the irreducibility of polynomials in $R[x]$, where R is a domain.

2.21. Theorem Let R be a domain and

$$f(x) = a_n x^n + \cdots + a_1 x + a_0$$

a polynomial in $R[x]$. Suppose there is a prime $p \in R$ such that
(a) $p \nmid a_n$,
(b) $p | a_i$, $i = 0, ..., n-1$,
(c) $p^2 \nmid a_0$.
Then $f(x)$ is irreducible in $R[x]$.

Proof Suppose $f(x) = g(x)h(x)$ for $g(x), h(x) \in R[x]$ where

$$g(x) = c_r x^r + \cdots + c_1 x + c_0$$

$$h(x) = d_s x^s + \cdots + d_1 x + d_0$$

with $c_i, d_j \in R$ and $r, s > 1$, $r + s = n$. Then by (b) and (c), $p | a_0 = c_0 d_0$ and hence p divides c_0 or d_0 but not both. Suppose $p | c_0$. By (a), we may let c_m be the first coefficient of $g(x)$ not divisible by p. But note that

$$a_m = c_0 d_m + c_1 d_{m-1} + \cdots + c_{m-1} d_1 + c_m d_0, \text{ where } p \nmid c_m d_0.$$

This implies $p \nmid a_m$, contradicting (b) because $m < n$. Hence $g(x)$ or $h(x)$ must be a unit of R. □

2.22. Corollary If p is a prime number, then the polynomial

$$f(x) = x^{p-1} + x^{p-2} + \cdots + x + 1$$

is irreducible in $\mathbb{Z}[x]$ and hence irreducible in $\mathbb{Q}[x]$.

Proof Note that $f(x) = \frac{x^p-1}{x-1}$. If we use the translation $x = X + 1$, then

$$f(X+1) + \frac{(X+1)^p - 1}{(X+1) - 1}$$

$$= \frac{1}{X}\left(X^p + \binom{p}{1}X^{p-1} + \binom{p}{2}X^{p-2} + \cdots + \binom{p}{p-1}X + 1 - 1\right)$$

$$= X^{p-1} + \binom{p}{1}X^{p-2} + \binom{p}{2}X^{p-3} + \cdots + p.$$

Now, using p as the prime needed in Theorem 2.21, we conclude that $f(x)$ is irreducible in $\mathbb{Z}[x]$ and hence irreducible in $\mathbb{Q}[x]$ by Proposition 2.18.

Exercises

1. Complete the proof of Proposition 2.2.
2. Complete the proof of Proposition 2.3.
3. Complete the proof of Proposition 2.8.
4. Let $R = K[t^2, t^3]$ be the subring generated by t^2 and t^3 in the polynomial ring $K[t]$ over a field K. Show that both t^2 and t^3 are irreducible in R but none is a prime. However $t^6 = t^2 t^2 t^2 = t^3 t^3$. (See also Chapter 3 (section 2, exercise 2) and Chapter 3 (section 3, Example (iii)).)
5. Show that the polynomial ring $R = K[x_1, x_2, ..., x_n, ...]$ in infinitely many variables over a field K is a UFD. (Hint: Any polynomial in R belongs to a polynomial ring in finitely many variables over K.)
6. Let R be a domain, $a, b \in R$ not all zero. Up to a unit multiple, define the greatest common divisor of a and b, denoted $\gcd(a,b)$, and the least common multiple of a and b (in case $a \neq 0$, $b \neq 0$), denoted $\mathrm{lcm}[a,b]$, as in $\mathbb{Z}$ (or as in $K[x]$ with K a field). (In a similar way, for $a_1, ..., a_n \in R$, $\gcd(a_1, ..., a_n)$ and $\mathrm{lcm}(a_1, ..., a_n)$ may be defined.)

   Show that the following statements are equivalent for a domain R in which factorization into irreducible elements is feasible.
   (a) R is a UFD.
   (b) Every irreducible element of R is a prime.
   (c) For every $a, b \in R$, not all zero, $\gcd(a,b)$ (or $\mathrm{lcd}[a,b]$ in case $a \neq 0$, $b \neq 0$) exists.
   (d) The intersection of two principal ideals of R is another principal ideal.
7. Let R be a UFD, $f, g \in R[x]$. Use Theorem 2.17 to show that if f, g

do not have common irreducible divisors in $R[x]$ then f, g do not have common irreducible divisors in $K[x]$ either, where $K = Q(R)$ is the field of fractions of R.

8. Let p be a prime number. Show that $x^n - p$ is irreducible in $\mathbb{Z}[x]$ and hence in $\mathbb{Q}[x]$.
9. Prove that $f = 11yx^8 + 3y^7x^5 + 9x^5 - 7y^7 - 21$ is irreducible in $\mathbb{Z}[x, y]$. (Hint: Consider f in $\mathbb{Z}[y][x]$.)

3. Field Extensions

The study of field extensions stems from the study of zeros of polynomials and the study of irreducibility of polynomials. Let K be a field and $f \in K[x_1, ..., x_n]$ a polynomial of degree ≥ 2. Then the property that f has or does not have a zero in K, and the property that f is reducible or irreducible over K, all depends on the ground field K, for instance, first consider the zeros of $x^2 - 1$, $x^2 - 2$ in $\mathbb{Q}$ and the zeros of $x^2 - 3$, $x^2 + 1$ in $\mathbb{R}$, and then consider the zeros of the given polynomials by extending $\mathbb{Q}$ to $\mathbb{R}$, $\mathbb{R}$ to $\mathbb{C}$. A full demonstration of this aspect is given in Chapter 4 and Chapter 5. In this section we focus on several fundamental topics concerning field extensions.

Let K, L be fields. If K is a subfield of L (including the case where $K \xrightarrow{\varphi} L$ is a ring monomorphism), then we call L an *extension field* of K, and from now on $K \subseteq L$ is referred to a *field extension*.

Let $K \subseteq L$ be a field extension and $S \subset L$ a subset of L. Consider the intersection

$$K(S) = \bigcap L_i$$

of all subfields in L containing S. Then it is an easy exercise to verify that

(a) $K(S)$ is the *smallest* subfield of L containing S, and
(b) $K(S) = Q(K[S])$, the field of fractions of $K[S]$ (hence $K(S)$ is also the smallest subfield of L containing $K[S]$).

In view of the above (a)–(b), we call $K(S)$ the subfield of L *generated by* S over K. If $S = \{s_1, ..., s_n\}$ is finite, then we write $K(S) = K(s_1, ..., s_n)$ and call it a *finitely generated* extension field of K. If S consists of a single element s, then $K(s)$ is called a *simple* extension field of K.

Splitting field

3.1. Definition Let K be a field, and let $f(x)$ be a polynomial in $K[x]$. If $K \subseteq L$ is a field extension such that $f(x)$ factors completely into linear factors over L, i.e., $f(x) = a\prod(x - \alpha_i)$ in $L[x]$, and $f(x)$ does not factor completely into linear factors over any proper subfield of L containing K, then L is called a *splitting field* of $f(x)$.

Let K be a field. To see the existence of a splitting field for an arbitrary $f(x) \in K[x]$, we start with an irreducible polynomial $p(x)$. Note that the quotient ring

$$L = \frac{K[x]}{\langle p(x)\rangle} = \left\{\overline{\psi(x)} = \sum \lambda_i \overline{x}^i \;\Big|\; \psi(x) \in K[x]\right\} = K[\overline{x}],$$

where $\overline{x}$ is the image of x in L, is a field, for, if $p(x) \nmid \psi(x)$ then $p(x)h(x) + \psi(x)g(x) = 1$ for some $h(x), g(x) \in K[x]$, and hence $\overline{\psi(x)}$ is invertible in L. Note that via the natural ring homomorphism $K[x] \to L$ we may write $K \subset L = K[\overline{x}]$. Thus,

$$K[x] \subset L[x] \text{ and consequently } p(\overline{x}) = 0.$$

It follows from the division algorithm that $p(x)$ is factorized in $L[x]$ as

$$p(x) = (x - \overline{x})p_1(x), \ p_1(x) \in L[x].$$

Now, since $K[x]$ is a UFD, an induction on the degree of polynomials, or a procedure of adding the zeros of each irreducible factor of $f(x)$ successively to the predecessor extension field, yields the following fact.

3.2. Theorem Let K be a field. Every $f(x) \in K[x]$ with $\deg f(x) = n > 0$ has a splitting field.

□

Example (i) The field $\mathbb{Q}(\sqrt{-3})$ serves as a splitting field for both $x^2 + 3$ and $x^3 + x^2 + 3x + 3$.

Remark Indeed, any splitting field of $f(x)$ is isomorphic to the one constructed before Theorem 3.2. The reader can refer to any textbook specifying field theory for a detailed proof.

Repeated zeros and separability

Let K be a field and let $f(x) \in K[x]$. In view of Theorem 3.2 we may always talk about the zeros of $f(x)$ in some extension field of K. Furthermore, we explore the following

Question When does $f(x)$ have no repeated zeros?

3.3. Proposition $f(x) \in K[x]$ has no repeated zeros if and only if $f(x)$ and $f'(x) = \frac{df(x)}{dx}$ are coprime, i.e., they do not have nonconstant common divisor.

Proof Over a splitting field E of $f(x)$, we have
$$f(x) = (x - \alpha_1)^{n_1} \cdots (x - \alpha_m)^{n_m}$$
where the α_i's are distinct. Then it is clear that $f(x)$ and $f'(x)$ have no nonconstant common divisor over E if and only if $n_i = 1$ for $i = 1, ..., m$. $\square$

3.4. Proposition Let E be a splitting field of $x^n - 1 = f(x) \in K[x]$, where $n \geq 1$. Suppose that char$K$ does not divide n. Then the following hold:
(i) $f(x)$ has exactly n distinct zeros (the nth roots of unity over K) in E.
(ii) Let
$$U_n = \left\{ \alpha \in E \mid f(\alpha) = 0 \right\}.$$
Then U_n is a cyclic multiplicative subgroup of $E^\times$.

Proof (i) By the assumption, this follows from Proposition 3.3.
(ii) That U_n forms a subgroup of $E^\times$ is clear. We show that U_n contains an element of order n. To this end, let
$$n = p_1^{e_1} \cdots p_s^{e_s}$$
be the factorization of n into primes, and let $q_i = \frac{n}{p_i}$ for $i = 1, ..., s$. Then, since the polynomial $x^{q_i} - 1$ has exactly q_i zeros in U_n, for each i, there is $\alpha_i \in U_n$ such that $\alpha_i^{q_i} \neq 1$. Set $\beta_i = \alpha_i^{n/p_i^{e_i}}$. Then $\beta_i^{p_i^{e_i-1}} \neq 1$ but $\beta_i^{p_i^{e_i}} = 1$. It follows that each β_i has order $p_i^{e_i}$. Since $p_1^{e_1}, ..., p_s^{e_s}$ are pairwise coprime, $\beta = \beta_1 \cdots \beta_s$ is the desired generator for U_n. $\square$

The last proposition makes the multiplicative structure of a finite field clear.

Preliminaries   21

3.5. Theorem Let K be a finite field. Then the multiplicative group $K^\times$ of K is cyclic.

Proof If $\operatorname{char} K = p > 0$, then $[K : \mathbb{Z}_p] = m$ for some m and hence $K^\times$ has $n = p^m - 1$ elements which are all zeros of $f(x) = x^n - 1 \in \mathbb{Z}_p[x]$. Since $p \nmid n$, Proposition 3.4 can be applied to this case. □

Since $K[x]$ is a UFD, the general discussion may be further reduced to irreducible elements.

3.6. Theorem Let K be a field and let $q(x) \in K[x]$ be irreducible.
(i) If $\operatorname{char} K = 0$, then $q(x)$ does not have repeated zeros.
(ii) If $\operatorname{char} K = p > 0$, then $q(x)$ has repeated zeros if and only if $q(x) = g(x^p)$ for some $g(x) \in K[x]$.

Proof We apply Proposition 3.3 to both cases.
(i) If $\operatorname{char} K = 0$, then since $q(x)$ is irreducible, we have $q'(x) \neq 0$ (otherwise $p(x)$ would be a constant), $\deg q'(x) < \deg q(x)$, and hence $q(x)$ and $q'(x)$ are coprime.
(ii) Suppose $\operatorname{char} K = p > 0$. Let $q(x) = a_n x^n + a_{n-1} x^{n-1} + \cdots + a_1 x + a_0$ with $a_n \neq 0$. Then $q'(x) = n a_n x^{n-1} + (n-1) a_{n-1} x^{n-2} + \cdots + a_1$ with $\deg q'(x) = n - 1 < \deg q(x) = n$. Thus,

$q(x)$ and $q'(x)$ have a nonconstant common divisor $\Leftrightarrow r a_r = 0$
$$\Leftrightarrow p | r, \text{ say } r = s_r p.$$

Consequently, $q(x)$ has repeated zeros if and only if $q(x) = a_{tp} x^{tp} + \cdots + a_{2p} x^{2p} + a_p x^p + a_0$. Therefore, $q(x) = g(x^p)$ where $g(y) = a_0 + a_p y + a_{2p} y^2 + \cdots + a_{tp} y^t$, as claimed. □

3.7. Corollary Let K be a finite field and let $q(x) \in K[x]$ be irreducible. Then $q(x)$ has no repeated zeros.

Proof Since K is finite, we know that $\operatorname{char} K = p > 0$ for some prime number p. Then $\mathbb{Z}_p$ is the prime field of K and K is a finite dimensional $\mathbb{Z}_p$-vector space, say $\dim_{\mathbb{Z}_p} K = n$. Hence K has p^n elements. Thus, the multiplicative group of K, which is $K^\times$, has order $p^n - 1$ and $\lambda^{p^n} = \lambda$ for all $\lambda \in K$. (We assumed that the reader is familiar with elementary group theory.) It follows that if $g(x^p) \in K[x]$, say $g(x^p) = a_0 + a_1 x^p + \cdots + a_n x^{np}$, then, after setting $a_i^{p^{n-1}} = b_i$, $i = 0, 1, ..., n$,

$$g(x^p) = a_0 + a_1 x^p + \cdots + a_n x^{np}$$
$$= b_0^p + b_1^p x^p + \cdots + b_n^p x^{np}$$
$$= (b_0 + b_1 x + \cdots + b_n x^n)^p,$$

which can never be irreducible. This shows that the irreducible $q(x)$ cannot have repeated zeros by Theorem 3.6. □

3.8. Definition If a polynomial $f(x) \in K[x]$ has no repeated zeros, then $f(x)$ is called a *separable polynomial* over K, and otherwise an *inseparable polynomial* over K. (See also Definition 3.11 below.)

Algebraic extension and primitive elements

We now start with a field extension $K \subseteq L$ and consider $\alpha \in L$. If there is some $f(x) \in K[x]$ such that $f(\alpha) = 0$, then we say that α is an *algebraic element* over K; otherwise, we say that α is a *transcendental element* over K. If every element of L is algebraic over K, then L is called an *algebraic extension field* of K, and we refer $K \subseteq L$ to an *algebraic field extension*. If L contains a transcendental element over K, then $K \subset L$ is referred to a *transcendental field extension*.

Let $K \subseteq L$ be a field extension. Then L is naturally viewed as a K-vector space. In the literature, the dimension $\dim_K L$ is also called the *degree* of L over K, denoted $[L : K]$.

Clearly, if a field extension $K \subseteq L$ has finite $[L : K]$, then L is algebraic over K. For instance, $[\mathbb{C} : \mathbb{R}] = 2$. If L contains a transcendental element over K, then $[L : K] = \infty$. It is known that e and π are transcendental over $\mathbb{Q}$. So $[\mathbb{R} : \mathbb{Q}] = \infty$. Another familiar transcendental extension is $K \subset K(x)$, where $K(x)$ is the field of fractions of the polynomial ring $K[x]$. Also, not every algebraic field extension is finite dimensional (exercise 4).

To understand the structure of a field extension $K \subseteq L$, simple extension plays a key role. Let $\alpha \in L$. Consider the subring $K[\alpha] \subseteq L$ and the ring homomorphism

$$\varphi: K[x] \longrightarrow K[\alpha]$$

$$g(x) \mapsto g(\alpha).$$

If α is a transcendental element over K, then

$$\text{Ker}\varphi = \{0\} \text{ and } K[x] \cong K[\alpha].$$

If α is algebraic over K, then $\ker\varphi \neq \{0\}$ and hence $\text{Ker}\varphi = \langle p(x) \rangle$ for some nonconstant $p(x) \in K[x]$ because $K[x]$ is a PID. We may assume that $p(x)$ is *monic*. It is a consequence of the division algorithm in $K[x]$ that $p(x)$ has the *smallest* positive degree among all polynomials in $\text{Ker}\varphi$. This leads to the following

3.9. Definition For an algebraic element α over K, the monic polynomial $p(x)$, which is the generator of $\ker\varphi$, is called the *minimal polynomial* of α over K.

3.10. Theorem Let $K \subseteq L$ be a field extension and $\alpha \in L$. If α is algebraic over K and $p(x)$ is its minimal polynomial, the following hold:
(i) $p(x)$ is irreducible and unique in $K[x]$.
(ii) $K[x]/\langle p(x) \rangle \cong K[\alpha]$ is a field containing K. Thus, $k[\alpha] = K(\alpha)$.
(iii) If $\deg p(x) = n$, then every element $\beta \in k(\alpha)$ has a unique expression

$$\beta = \lambda_{n-1}\alpha^{n-1} + \lambda_{n-1}\alpha^{n-2} + \cdots + \lambda_1\alpha + \lambda_0, \ \lambda_i \in K.$$

Thus, $\{\alpha^{n-1}, ..., \alpha, 1\}$ forms a K-basis for $K(\alpha)$, $[K(\alpha) : K] = n$. Consequently, $K(\alpha)$ is a simple algebraic extension field of K.

Proof Using division algorithm by $p(x)$ in $K[x]$, all conclusions are easy exercises. □

Later in exercise 2 the reader will be asked to show that if $\alpha_1, ..., \alpha_m \in L$ are finitely many algebraic elements over K, then $K \subseteq K(\alpha_1, ..., \alpha_m)$ is an algebraic field extension and $[F : K] < \infty$. When K plays the role as in the case of Theorem 3.6(i) and Corollary 3.7, our next goal is to show that the finitely generated algebraic field extension $K(\alpha_1, ..., \alpha_m)$ is actually a simple extension. But first, we need the notion of a separable extension.

3.11. Definition (i) Let $K \subseteq L$ be a field extension and let $\alpha \in L$ be an algebraic element over K. If the minimal polynomial $p(x)$ of α over K is separable in the sense of Definition 3.8, then α is said to be *separable* over K; otherwise α is *inseparable* over K.

(ii) Let $K \subseteq L$ be an algebraic field extension. If every element of L is separable over K, then L is said to be *separable* over K; otherwise L is *inseparable* over K.

By Theorem 3.6 and Corollary 3.7, inseparable field extension is, indeed, quite rare.

3.12. Theorem Let $K \subseteq F = K(\alpha_1, \alpha_2, ..., \alpha_m)$ be a finitely generated algebraic field extension. Suppose that $\alpha_2, ..., \alpha_m$ are separable over K. Then $F = K(\vartheta)$ for some $\vartheta \in F$.

Proof If K is finite then so is F (by Exercise 2), and the conclusion follows from Theorem 3.5.

Suppose that K is infinite. We consider only the case where $F = K(\alpha, \beta)$ with β separable over K since the general conclusion may be obtained by an induction.

Let L be a field over which the minimal polynomial $p(x)$ of α and the minimal polynomial $q(x)$ of β are factorized as

$$p(x) = \prod_{i=1}^{n}(x - \alpha_i), \qquad q(x) = \prod_{j=1}^{m}(x - \beta_j),$$

where $\alpha_1, \alpha_2, ..., \alpha_n, \beta_1, ..., \beta_m \in L$, and $\alpha_1 = \alpha$, $\beta_1 = \beta$. (The existence of L is guaranteed by Theorem 3.2.) By the assumption, $\beta_1, ..., \beta_m$ are distinct. Thus, the equations

$$\alpha_i - \alpha_1 = \lambda_{ik}(\beta_1 - \beta_k), \ k \neq 1,$$

have only finitely many solutions $\lambda_{ik} \in K$. Hence, there exists $c \in K$ such that

$$\alpha_i - \alpha_1 \neq c(\beta_1 - \beta_k), \ 1 \leq i \leq n, \ 2 \leq k \leq m.$$

Set $\vartheta = \alpha + c\beta$. Then clearly $K(\vartheta) \subseteq F$. Below we show that $\beta \in K(\vartheta)$ and then it follows that $F = K(\vartheta)$.

Note that $\alpha = \vartheta - c\beta$. We have $p(\vartheta - c\beta) = p(\alpha) = 0$. Consider the polynomial $r(x) = p(\vartheta - cx) \in K(\vartheta)[x]$. Then, by the choice of c, β is the only common zero of $q(x)$ and $r(x)$ in F. This shows that the minimal polynomial of β in $K(\vartheta)[x]$ is of the form $t - \mu$ for some $\mu \in K(\vartheta)$. Therefore, $\beta = \mu \in K(\vartheta)$ as expected. $\square$

3.13. Definition The element ϑ that appears in Theorem 3.12 is called a *primitive element* of F.

Example (ii) $K = \mathbb{Q}(\sqrt{2}, \sqrt{3}) = \mathbb{Q}(\sqrt{2} + \sqrt{3})$.

Let F be a field. If every nonconstant polynomial $f(x) \in F[x]$ splits in F, i.e., $f(x) = \prod_{i=1}^{n} \lambda(x - \lambda_i)$, $\lambda, \lambda_i \in F$, then F is said to be *algebraically closed*. Clearly, if F is algebraically closed, then there is no proper algebraic extension of F. For instance, the field $\mathbb{C}$ of complex numbers is algebraically closed (this is also known as the content of the fundamental theorem of algebra). Without proof we mention the following theorem (the reader is referred to any textbook specializing field theory for the classical proof given by Emil Artin).

Theorem Let K be a field. Then there is an extension field L of K that is algebraically closed.

Lüroth's theorem

Within the context of Theorem 3.6(i), Corollary 3.7 and Theorem 3.12, it is easy to see that if $K \subseteq L = K(\vartheta)$ is a simple field extension, then any intermediate field extension F of K with $K \subsetneq F \subseteq L$ is a simple extension. The final part of this section deals with a similar situation on simple transcendental field extension.

Let K be a field and x a transcendental element over K. Given coprime polynomials $u(x), v(x) \in K[x]$, consider $h = \frac{u(x)}{v(x)} \in K(x)$ and the simple extension $K \subset K(h)$. Set

$$hv(t) - u(t) = q(t) \in K(h)[t],$$

where $K(h)[t]$ is the polynomial ring in t over $K(h)$.

3.14. Lemma With notation as above, the following hold:
(i) h is transcendental over K.
(ii) $q(t)$ is irreducible in $K(h)[t]$.
(iii) $[K(x) : K(h)] = \deg q(t) = \max\{\deg u(x), \deg v(x)\}$.

Proof (i) Exercise.
(ii) Note that $q(t)$ is linear with respect to h in the polynomial ring $K[h,t]$ which is a UFD. Hence $q(t)$ is irreducible in $K[h,t]$, for $u(x)$ and $v(x)$ are coprime by the assumption. It follows from Proposition 2.18 that $q(t)$ is irreducible in $K(h)[t]$.
(iii) By the construction of $q(t)$, $q(x) = 0$. It follows from part (ii) that $q(t)$ (assuming monic) is the minimal polynomial of x over $K(h)$. Thus, $[K(x) : K(h)] = \deg q(t) = \max\{\deg u(x), \deg v(x)\}$, as desired. □

3.15. Corollary (i) Let E be any intermediate extension field of K with $K \subsetneq E \subseteq K(x)$. Then $[K(x) : E] < \infty$.
(ii) Every automorphism of the ring $K(x)$ which is K-linear is given by

$$x \mapsto \frac{ax+b}{cx+d}, \quad a,b,c,d \in K, \ ad - bc \neq 0.$$

Proof Exercise. □

3.16. Theorem (Lüroth) Let K be a field and x a transcendental element over K. Let E be an intermediate extension field of K with $K \subsetneq E \subseteq K(x)$. Then $E = K(y)$ for some $y \in K(x)$ (hence $E \cong K(x)$) and $[K(x) : E] < \infty$.

Proof By Corollary 3.15, $[K(x) : E] < \infty$. Let $p(t) \in E[t]$ be the minimal polynomial of x over E, say

$$p(t) = t^n + r_{n-1}t^{n-1} + \cdots + r_0, \quad r_i \in E.$$

If we multiply $p(t)$ by the least common multiple, say s, of the denominators of r_i's, the obtained polynomial

(1) $\qquad f(x,t) = sp(t) = s_n t^n + s_{n-1}t^{n-1} + \cdots + s_0, \quad s_i \in K[x],$

is primitive in $K[x][t]$ with respect to t (check it!). Write $\deg_t f(x,t)$ for the degree of $f(x,t)$ in t. Then

$$n = \deg_t f(x,t) = \deg p(t) = [K(x) : E].$$

Note that $s_n = s$ and all $\frac{s_i}{s_n} \in E$. As x is transcendental over K, there is at least one $\frac{s_i}{s_n} \in E - K$. Set $h = \frac{u(x)}{v(x)} = \frac{s_i}{s_n}$ for convenience, where $u(x)$ and $v(x)$ are coprime in $K[x]$. Then

$$q(t) = hv(t) - u(t)$$

is irreducible in $K(h)[t]$ and

(2) $\qquad [K(x):K(h)] = \deg q(t) = \max\{\deg u(x), \deg v(x)\}$

by Lemma 3.14. Since $K \subsetneq K(h) \subseteq E \subseteq K(x)$, we complete the proof by having the equality

$$[K(x):E] = [K(x):K(h)].$$

To this end, note that $q(x) = 0$ and $q(t) \in E[t]$. Hence $q(t) = p(t)p_1(t)$ with $p_1(t) \in E[t]$, for $p(t)$ is the minimal polynomial of x over E. Thus, by formula (1),

$$u(x)v(t) - v(x)u(t) = v(x)p(t)p_1(t)$$

$$= \left(\frac{v(x)}{s}p_1(t)\right)f(x,t).$$

But $f(x,t)$ is primitive in $K[x][t]$ with respect to t. It follows from Theorem 2.17(ii) that

(3) $\qquad u(x)v(t) - v(x)u(t) = df(x,t), \quad d \in K[x][t].$

Suppose $\deg_x f(x,t) = m$. Then $\max\{\deg u(x), \deg v(x)\} \leq m$ by formula (1). So the above formula (3) implies that

(4) $\qquad \deg_x(u(x)v(t) - v(x)u(t)) = m$

and d is a constant. Note that $u(x)v(t) - v(x)u(t)$ is antisymmetric in x and t. Therefore, (2) + (4) yields

$$[K(x):K(h)] = \max\{\deg u(x), \deg v(x)\} = m$$

$$= \deg_x(u(x)v(t) - v(x)u(t))$$

$$= \deg_t(u(x)v(t) - v(x)u(t))$$

$$= \deg_t f(x,t)$$

$$= n = [K(x):E],$$

as desired. $\qquad \square$

Exercises

1. Let $K \subseteq L$ be a field extension and $\alpha_1, ..., \alpha_m \in L$. Show that if $\alpha_1, ..., \alpha_m$ are algebraic over K, then $K \subseteq F = K(\alpha_1, ..., \alpha_m)$ is an algebraic field extension and $[F : K]$ is finite. (Hint: Note that $F = K(\alpha_1)(\alpha_2) \cdots (\alpha_m)$ and use Theorem 3.10(iii).)
2. Let $K \subseteq L \subseteq E$ be a tower of algebraic field extension, i.e., L is algebraic over K and E is algebraic over L. Use exercise 1 to show that E is also algebraic over K. Moreover, show that if $[L : K] < \infty$ and $[E : L] < \infty$, then $[E : K] = [L : K][E : L]$. (Hint: If $\alpha \in E$ and $\lambda_n \alpha^n + \cdots + \lambda_1 \alpha + \lambda_0 = 0$ for $\lambda_i \in L$, then consider $K \subseteq K(\lambda_n, ..., \lambda_0) \subseteq K(\lambda_n, ..., \lambda_0)(\alpha)$.)
3. Use Theorem 3.10(iii) to show that if $K \subseteq L$ is a field extension, then all elements of L which are algebraic over K form a subfield $\widehat{K}$ of L containing K. $\widehat{K}$ is called the *algebraic closure* of K in L. (Hint: For $\alpha, \beta \in L$, algebraic over K, consider $K \subseteq K[\alpha] \subseteq k[\alpha][\beta]$.)
4. Let F be the subfield of $\mathbb{C}$ consisting of all algebraic elements over $\mathbb{Q}$. Use (section 2, exercise 8) to show that $[F : \mathbb{Q}] = \infty$.
5. Show that if K is an algebraically closed field, then K is infinite (or equivalently, that a finite field cannot be algebraically closed). (Hint: If $F = \{a_1, ..., a_n\}$ is a finite field, consider the polynomial $p(x) = \prod_{i=1}^{n}(x - a_i) + 1$ in $F[x]$.)
6. Let $d \in \mathbb{Z}$ be square-free. Then every element $\alpha \in \mathbb{Q}(\sqrt{d})$ is of the form $\alpha = r + s\sqrt{d}$, where $r, s \in \mathbb{Q}$. Show that α has the minimal polynomial

$$p_\alpha(x) = x^2 - 2rx + (r^2 - s^2 d).$$

7. Let $F = \mathbb{Q}(\sqrt{2}, \sqrt[3]{5})$. Find a primitive element for F.
8. Prove Lemma 3.14(i).
9. Complete the proof of Corollary 3.15.

4. Symmetric Polynomials

Let R be a ring and $R[x_1, ..., x_n]$ the polynomial ring in variables $x_1, ..., x_n$ over R. Put

$$\mathbb{N}^n = \left\{ \alpha = (\alpha_1, ..., \alpha_n) \,\middle|\, \alpha_i \in \mathbb{N} \right\}.$$

Then every element $f(x_1, ..., x_n) \in R[x_1, ..., x_n]$ has a *unique* expression

(*) $\quad f(x_1, ..., x_n) = \sum_\alpha c_\alpha x_1^{\alpha_1} \cdots x_n^{\alpha_n}, \ \alpha = (\alpha_1, ..., \alpha_n) \in \mathbb{N}^n, \ c_\alpha \in R.$

Let S_n denote the permutation group of $\{1, 2, ..., n\}$. A polynomial $f(x_1, ..., x_n) \in R[x_1, ..., x_n]$ is said to be *symmetric* if

$$f(x_1, ..., x_n) = f(x_{\pi(1)}, x_{\pi(2)}, ..., x_{\pi(n)}), \text{ for all } \pi \in S_n.$$

For example, $x_1^2 + x_2^2 + x_3^2$, $(x_1 + x_2 + x_3 + x_4)(x_1 x_2 x_3 x_4)^3$.

Important symmetric polynomials are those *elementary symmetric polynomials*:

$$s_1(x_1, ..., x_n) = x_1 + x_2 + \cdots + x_n$$

$$s_2(x_1, ..., x_n) = x_1 x_2 + x_1 x_3 + \cdots + x_1 x_n$$

$$+ x_2 x_3 + x_2 x_4 + \cdots + x_2 x_n$$

$$\vdots$$

$$+ x_{n-2} x_{n-1} + x_{n-2} x_n$$

$$+ x_{n-1} x_n$$

$$\vdots$$

$$s_k(x_1, ..., x_n) = \sum_{1 \leq i_1 < i_2 < \cdots < i_k \leq n} x_{i_1} x_{i_2} \cdots x_{i_k}$$

$$\vdots$$

$$s_n(x_1, ..., x_n) = x_1 x_2 \cdots x_n.$$

Let $R[s_1, ..., s_n]$ be the subring of $R[x_1, ..., x_n]$ generated by R and $\{s_1, s_2, ..., s_n\}$. Then it is clear that every $g(s_1, s_2, ..., s_n) \in R[s_1, ..., s_n]$ is a symmetric polynomial. The next theorem, due to Newton, shows that the converse is also true.

4.1. Theorem If $f = f(x_1,...,x_n)$ is a symmetric polynomial in $R[x_1,...,x_n]$, then $f(x_1,...,x_n) \in R[s_1,...,s_n]$.

Proof To reduce $f = f(x_1,...,x_n) = \sum_\alpha c_\alpha x_1^{\alpha_1} \cdots x_n^{\alpha_n}$ into a polynomial in elementary symmetric polynomials, in view of previous (*) we order the set of monomials

$$\left\{ x_1^{\alpha_1} x_2^{\alpha_2} \cdots x_n^{\alpha_n} \;\Big|\; (\alpha_1,...,\alpha_n) \in \mathbb{N}^n \right\}$$

by the lexicographic ordering:

$$x_1^{\alpha_1} \cdots x_n^{\alpha_n} \prec_{lex} x_1^{\beta_1} \cdots x_n^{\beta_n}$$

if and only if

$$\alpha_1 = \beta_1,\; \alpha_2 = \beta_2,...,\alpha_{s-1} = \alpha_{s-1} \text{ while } \alpha_s < \beta_s \text{ for some } s \leq n.$$

Thus, the terms of f are ordered lexicographically (note that $\prec_{lex}$ is a total ordering), and we may assume that the leading monomial of f is $x_1^{\alpha_1} x_2^{\alpha_2} \ldots x_n^{\alpha_n}$.

Since f is symmetric, $x_{\pi(1)}^{\alpha_1} x_{\pi(2)}^{\alpha_2} \cdots x_{\pi(n)}^{\alpha_n}$ occurs in f for every $\pi \in S_n$. It follows that the leading monomial of f has the property that $\alpha_1 \geq \alpha_2 \geq \cdots \geq \alpha_n$. For example, the leading monomial of

$$s_1^{k_1} s_2^{k_2} \cdots s_n^{k_n} = (x_1 + \cdots + x_n)^{k_1} \cdots (x_1 \cdots x_n)^{k_n}$$

is

$$x_1^{k_1+\cdots+k_n} x_2^{k_2+\cdots+k_n} \cdots x_n^{k_n}.$$

By choosing $k_1 = \alpha_1 - \alpha_2,..., k_{n-1} = \alpha_{n-1} - \alpha_n, k_n = \alpha_n$, we can make this the same as the leading monomial of f. Suppose that the leading coefficient of f is c, then $f - cs_1^{k_1} s_2^{k_2} \cdots s_n^{k_n}$ has a lexicographic leading term

$$dx_1^{\beta_1} x_2^{\beta_2} \cdots x_n^{\beta_n}, \quad \beta_1 \geq \beta_2 \geq \cdots \geq \beta_n$$

which comes after $cx_1^{\alpha_1} x_2^{\alpha_2} \ldots x_n^{\alpha_n}$ in the ordering. Since only a finite number of monomials $x_1^{\gamma_1} x_2^{\gamma_2} \cdots x_n^{\gamma_n}$ in f satisfying $\gamma_1 \geq \gamma_2 \geq \cdots \geq \gamma_n$ follow $x_1^{\alpha_1} x_2^{\alpha_2} \ldots x_n^{\alpha_n}$ lexicographically, a finite number of repetitions of the above process reduce f to a polynomial in $s_1,...,s_n$. $\square$

Example (i) The symmetric polynomial

$$f = x_1^2 x_2 + x_1^2 x_3 + x_1 x_2^2 + x_1 x_3^2 + x_2^2 x_3 + x_2 x_3^2$$

is written lexicographically. And by the method given in the proof we may derive that $f = s_1s_2 + 3s_3$. Similarly, $(x_1+x_2)(x_1+x_3)(x_2+x_3) = s_1s_2 - s_3$.

An application of symmetric polynomials to field extension is given as follows.

If $K \subseteq L$ is a field extension, $a_n x^n + a_{n-1} x^{n-1} + \cdots + a_0 = f(x) \in K[x]$ with $\deg f(x) = n$, and $f(r_i) = 0$ with $r_1, ..., r_n \in L$, then, $f(x)$ factors in $L[x]$ as

$$f(x) = a_n(x - r_1)(x - r_2) \cdots (x - r_n)$$

$$= a_n(x^n + c_1 x^{n-1} + c_2 x^{n-2} + \cdots + c_n)$$

where $c_i = (-1)^i s_i(r_1, r_2, ..., r_n)$, $i = 1, ..., n$. After comparing coefficients of both sides, we have

$$(-1)^i a_n s_i(r_1, r_2, ..., r_n) = a_{n-i} \in K, \quad i = 1, ..., n.$$

4.2. Corollary Let $K \subseteq L$ be a field extension, $a_n x^n + a_{n-1} x^{n-1} + \cdots + a_0 = f \in K[x]$ with $\deg f = n$, and $f(r_i) = 0$ with $r_1, ..., r_n \in L$. If $h(x_1, ..., x_n) \in K[x_1, ..., x_n]$ is a symmetric polynomial, then $h(r_1, r_2, ..., r_n) \in K$, i.e., $\{r_1, ..., r_n\}$ defines a function

$$K[s_1, ..., s_n] \longrightarrow K$$

$$h \mapsto h(r_1, ..., r_n)$$

□

Example (ii) Suppose that r_1, r_2, r_3 are the zeros of $f(x) = x^3 + x^2 - x + 1$ in $\mathbb{C}$. Find $r_1^2 + r_2^2 + r_3^2$ and $r_1^3 + r_2^3 + r_3^3$.

Solution Since $f(x) = (x - r_1)(x - r_2)(x - r_3)$, it follows that

$$r_1 + r_2 + r_3 = -1,$$

$$r_1 r_2 + r_1 r_3 + r_2 r_3 = -1,$$

$$r_1 r_2 r_3 = -1.$$

From $(r_1 + r_2 + r_3)^2 = r_1^2 + r_2^2 + r_3^2 + 2(r_1 r_2 + r_1 r_3 + r_2 r_3)$ we derive that $r_1^2 + r_2^2 + r_3^2 = 1 + 2 = 3$; and from $f(x_i) = 0$, $i = 1, 2, 3$, we derive that $r_1^3 + r_2^3 + r_3^3 = -(r_1^2 + r_2^2 + r_3^2) + (r_1 + r_2 + r_3) - 3 = -7$.

More generally, the following recurrence relations, called *Newton's formulas*, can be used to establish formulas for $p_i = (-1)^i(x_1^i + x_2^i + \cdots + x_n^i)$, $i \geq 1$, in terms of $s_1, s_2, \ldots, s_n$.

$$p_1 + s_1 = 0,$$

$$p_2 + s_1 p_1 + 2s_2 = 0,$$

$$p_3 + s_1 p_2 + s_2 p_1 + 3s_3 = 0,$$

$$\ldots$$

$$p_n + s_1 p_{n-1} s_2 p_{n-2} + \cdots + s_{n-1} p_1 + n s_n = 0.$$

We close with an application to polynomial building.

Example (iii) Let r_1, r_2, r_3 be the zeros of $f(x) = x^3 - x + 2$ in $\mathbb{C}$. Find the polynomial $g(x)$ that has zeros r_1^2, r_2^2, r_3^2.

Solution Suppose the desired polynomial is of the form $g(x) = x^3 + Ax^2 + Bx + C$. Then

$$A = -(r_1^2 + r_2^2 + r_3^2) = -p_2(r_1, r_2, r_3)$$

$$= -s_1(r_1, r_2, r_3)^2 + 2s_2(r_1, r_2, r_3)$$

$$= 0 + 2(-1) = -2,$$

$$B = r_1^2 r_2^2 + r_1^2 r_3^2 + r_2^2 r_3^2 = s_2(r_1, r_2, r_3)^2 - 2s_1(r_1, r_2, r_3)s_3(r_1, r_2, r_3)$$

$$= (-1)^2 - 2(0)(-2) = 1,$$

$$C = -r_1^2 r_2^2 r_3^2 = -s_3(r_1, r_2, r_3)^2 = -4.$$

Hence $g(x) = x^3 - 2x^2 + x - 4$.

Exercises

1. Express the product $(x_1^2 + x_2^2)(x_1^2 + x_3^2)(x_2^2 + x_3^2)$ in terms of s_1, s_2, s_3.
2. Let r_1, r_2, r_3 be the zeros of $f(x) = x^3 - 6x + 11 - 6$ in $\mathbb{C}$. Determine the polynomial $g(x)$ that has zeros $r_1^2 + r_2^2$, $r_1^2 + r_3^2$, $r_2^2 + r_3^2$.
3. Let r_1, r_2, r_3, r_4 be the zeros of $f(x) = a_4 x^4 + a_3 x^3 + a_2 x^2 + a_1 x + a_0$ in

$\mathbb{C}$, where $a_i \in \mathbb{Q}$. Suppose $a_4 = -5$ and the elementary polynomials in r_1, r_2, r_3, r_4 are $s_1 = \frac{3}{5}$, $s_2 = 16$, $s_3 = -8$, $s_4 = -\frac{1}{10}$. Find a_3, a_2, a_1, a_0.

4. Let R be a ring. A polynomial belonging to $R[x_1, .., x_n]$ is said to be *antisymmetric* if it is invariant under even permutations of the variables, but changes sign under odd permutations. Let

$$\Delta = \prod_{i<j}(x_i - x_j).$$

Show that
(a) Δ is antisymmetric, and
(b) if $2r = 0$ implies $r = 0$ for $r \in R$, then any antisymmetric polynomial f is expressible as a polynomial in the elementary symmetric polynomials, together with Δ. (Hint: Note that $f(x_1, x_2, x_3, ..., x_n) = -f(x_2, x_1, x_3, ..., x_n)$, $2f(x_1, x_1, x_3, ..., x_n) = 0$. Thus, f vanishes when $x_1 = x_2$. So a division on f by $x_1 - x_2$ in $R[x_2, ..., x_n][x_1]$ yields $(x_1 - x_2)|f$. Similarly, $(x_1 - x_i)|f$, $i = 3, ..., n$. Writing $f = \prod_{i \neq 1}(x_1 - x_i)f_1$, where $f_1 \in R[x_2, ..., x_n]$ and is antisymmetric. Now an induction on n finishes the proof.)

5. Trace and Norm

Throughout this section we let $K \subsetneq L$ be a *simple algebraic field extension*, that is, $L = K(\vartheta)$, $\vartheta \in L$. If $p(x) \in K[x]$ is the minimal polynomial of ϑ over K, we may set a tower of field extensions

$$K \subset L \subset E$$

such that E contains *all distinct zeros* of $p(x)$, say $\vartheta_1 = \vartheta, \vartheta_2, ..., \vartheta_m$, where $m \leq n = [L:K] = \deg p(x)$, that is, E contains the splitting field of $p(x)$.

5.1. Proposition With notation as above, there are exactly m distinct ring monomorphisms $L \to E$ that are K-linear. Moreover each K-linear monomorphism $L \to E$ is given by $\vartheta \to \vartheta_i$, $1 \leq i \leq m$.

Proof If $\sigma: L \to E$ is a monomorphism as described, then $0 = \sigma(p(\vartheta)) = p(\sigma(\vartheta))$, i.e., $\sigma(\vartheta)$ is a zero of $p(x)$. Note that the elements of L are of the form $\sum \lambda_j \vartheta^j$. It follows that if $L \xrightarrow{\sigma_1} E$ and $L \xrightarrow{\sigma_2} E$ are two K-linear ring

monomorphisms such that $\sigma_1(\vartheta) = \sigma_2(\vartheta)$, then $\sigma_1 = \sigma_2$.
Conversely, each ϑ_i defines a desired monomorphism

$$\sigma_i : \quad L \longrightarrow E$$

$$\sum \lambda_j \vartheta^j \mapsto \sum \lambda_j \vartheta_i^j$$

because all ϑ_i's have the same minimal polynomial $p(x)$. $\square$

With the help of Proposition 5.1 we may determine, for every $\alpha \in L$, the minimal polynomial $p_\alpha(x)$ of α over K and the splitting field of $p_\alpha(x)$. To see this, let $\sigma_1, ..., \sigma_m$ be all distinct monomorphisms $L \to E$ defined by $\sigma_i(\vartheta) = \vartheta_i$, $i = 1, ..., m$. Suppose that each ϑ_i has multiplicity $e_i \geq 1$, that is, $p(x) = \prod_{i=1}^{m}(x - \vartheta_i)^{e_i}$ in $E[x]$. Then

$$e_1 + e_2 + \cdots + e_m = n = \deg p(x),$$

and each $\alpha \in L = K(\vartheta)$ is associated to a monic polynomial in $E[x]$, that is,

$$f_\alpha(x) = \prod_{i=1}^{m}(x - \sigma_i(\alpha))^{e_i}.$$

For convenience, we call $f_\alpha(x)$ the *total polynomial* of α.

5.2. Proposition Let $K \subseteq L = K(\vartheta) \subset E$ be as above. For any $\alpha \in L = K(\vartheta)$, the following hold:
(i) $f_\alpha(x) \in K[x]$.
(ii) Let $p_\alpha(x) \in K[x]$ be the minimal polynomial of α over K. Then $f_\alpha(x) = p_\alpha(x)^s$ for some $s \geq 1$.
(iii) E contains the splitting field of the minimal polynomial $p_\alpha(x)$ of α over K.

Proof (i) Since $\alpha = r(\vartheta) = \sum_{i=1}^{n-1} \lambda_i \vartheta^i$, where $r(x) = \sum_{i=1}^{n-1} \lambda_i x^i \in K[x]$, we have

$$f_\alpha(x) = \prod_{i=1}^{m}(x - \sigma_i(r(\vartheta)))^{e_i}$$

$$= \prod_{i=1}^{m}(x - r(\vartheta_i))^{e_i}.$$

Note that all $\lambda_i \in K$. After expanding the latter product we see that the coefficients of $f_\alpha(x)$ are given by symmetric polynomials in the n zeros of $p(x)$. By Corollary 4.2, $f_\alpha(x) \in K[x]$.

(ii) By part (i), $f_\alpha(x) \in K[x]$ and $f_\alpha(\alpha) = 0$. It follows that $f_\alpha(x) = p_\alpha(x)^s h(x)$, where $h(x) \in K[x]$ and $p_\alpha(x)$, $h(x)$ are coprime and both are monic. If $h(x)$ is not a constant, then some $\sigma_i(\alpha)$ is a zero of $h(x)$. Let $\alpha = r(\vartheta) = \sum_{i=1}^{n-1} \lambda_i \vartheta^i$ with $r(x) = \sum_{i=1}^{n-1} \lambda_i x^i \in K[x]$. Then $\sigma_i(\alpha) = r(\vartheta_i)$ and $h(\sigma_i(\alpha)) = h(r(\vartheta_i)) = 0$. Set $g(x) = h(r(x)) \in K[x]$. Then $g(\vartheta_i) = h(r(\vartheta_i)) = 0$ implies $p(x)|g(x)$, for $p(x)$ is the minimal polynomial of ϑ and hence the minimal polynomial of each ϑ_i. It follows that $0 = g(\vartheta) = h(r(\vartheta)) = h(\alpha)$ and $p_\alpha(x)|h(x)$, a contradiction. This shows that $h(x)$ is a constant and $h(x) = 1$ because it is monic. Thus, $f_\alpha(x) = p_\alpha(x)^s$.

(iii) By parts (i) and (ii), $f_\alpha(x) \in K[x]$ and $f_\alpha(x) = \prod_{i=1}^{m}(x - \sigma_i(\alpha))^{e_i} = p_\alpha(x)^s$ for some $s \geq 1$. So $p_\alpha(x)$ factors into linear divisors over E. □

We now introduce two functions on L that will play important roles in Chapter 3 section 3 and throughout Chapter 4.

Let $\alpha \in L = K(\vartheta)$. By Proposition 5.2(i), the total polynomial $f_\alpha(x)$ of α belongs to $K[x]$. By the definition, $f_\alpha(x) = \prod_{i=1}^{m}(x - \sigma_i(\alpha))^{e_i}$, where $e_1 + \cdots + e_m = n = [L : K] = \deg p(x)$. If we check the expanded expression of $f_\alpha(x)$, then $f_\alpha(x) = x^n + c_{n-1}x^{n-1} + \cdots + c_1 x + c_0$ with $c_{n-1} = -\sum_{i=1}^{m} e_i \sigma_i(\alpha)$ and $c_0 = (-1)^n \prod_{i=1}^{m} \sigma_i(\alpha)^{e_i}$. Thus, we have obtained two well-defined functions:

$$T_{L/K} : L \longrightarrow K$$

$$\alpha \mapsto \sum_{i=1}^{m} e_i \sigma_i(\alpha)$$

$$N_{L/K} : L \longrightarrow K$$

$$\alpha \mapsto \prod_{i=1}^{m} \sigma_i(\alpha)^{e_i}$$

5.3. Definition For $\alpha \in L$, $T_{L/K}(\alpha)$ is called the *trace* of α in K and $N_{L/K}(\alpha)$ is called the *norm* of α in K.

5.4. Proposition For $\alpha, \beta \in L$, $\lambda, \mu \in K$, the following hold:
(i) $T_{L/K}(\lambda \alpha + \mu \beta) = \lambda T_{L/K}(\alpha) + \mu T_{L/K}(\beta)$.

(ii) $T_{L/K}(\alpha\beta) = T_{K/L}(\beta\alpha)$.
(iii) $N_{L/K}(\alpha\beta) = N_{L/K}(\alpha)N_{L/K}(\beta)$.

Proof Exercise. □

Knowledge on bilinear forms needed by the next theorem and Chapter 3 Theorem 3.2 is given as an appendix at the end of this section.

Viewing L as an n-dimensional K-vector space, Proposition 5.4 enables us to define a symmetric bilinear form on L:

$$L \times L \longrightarrow K$$

$$(\alpha, \beta) \mapsto T_{L/K}(\alpha\beta)$$

5.5. Theorem For $K \subseteq L = K(\vartheta)$ with $[L:K] = n$, if the minimal polynomial $p(x)$ of ϑ over K has n *distinct* zeros $\vartheta_1 = \vartheta, \vartheta_2, ..., \vartheta_n$, then the bilinear form defined above is nondegenerate.

Proof Set $r(k) = (\vartheta_1^k, ..., \vartheta_n^k)$, $k = 0, ..., n-1$, and write V for the Vandermonde matrix

$$V = \begin{pmatrix} r(0) \\ r(1) \\ r(2) \\ \vdots \\ r(n-1) \end{pmatrix} = \begin{pmatrix} 1 & 1 & \cdots & 1 \\ \vartheta_1 & \vartheta_2 & \cdots & \vartheta_n \\ \vartheta_1^2 & \vartheta_2^2 & \cdots & \vartheta_n^2 \\ \vdots & \vdots & \cdots & \vdots \\ \vartheta_1^{n-1} & \vartheta_2^{n-1} & \cdots & \vartheta_n^{n-1} \end{pmatrix}.$$

Let $\sigma_1, ..., \sigma_n$ be all the n distinct K-linear monomorphisms from L to E as described in Proposition 5.1 such that $\sigma_i(\vartheta) = \vartheta_i$, $i = 1, ..., n$. If we consider the standard K-basis $\{1, \vartheta, ..., \vartheta^{n-1}\}$ of L, then since

$$T_{L/K}(\vartheta^k\vartheta^j) = \sum_{i=1}^n \sigma_i(\vartheta^k\vartheta^j) = \sum_{i=1}^n \vartheta_i^k\vartheta_i^j = \vartheta(k)(\vartheta(j))^t,$$

the matrix of the bilinear form is given by

$$\left(T_{L/K}(\vartheta^k\vartheta^j)\right) = VV^t$$

and hence

$$\det\left(T_{L/K}(\vartheta^k\vartheta^j)\right) = (\det(V))^2.$$

Since all the ϑ_i are distinct, $\det(V) = \prod_{i<j}(\vartheta_i - \vartheta_j) \neq 0$. This shows that the bilinear form is nondegenerate. □

5.6. Corollary If $K \subseteq L$ is a finite dimensional separable field extension, then Theorem 5.5 hold.

$\square$

Appendix. Bilinear forms

Let U and V be two vector spaces over a field K, $U \times V = \{(u,v) \mid u \in U, v \in V\}$ the Cartesian product of U and V. A *bilinear form* on $U \times V$ is a mapping

$$< \ , \ >: \ U \times V \longrightarrow \ K$$

$$(u,v) \ \mapsto \ <u,v>$$

satisfying

$$<\lambda u_1 + \mu u_2, v> = \lambda <u_1, v> + \mu <u_2, v>$$

$$<u, \lambda v_1 + \mu v_2> = \lambda <u, v_1> + \mu <u, v_2>$$

for all $u_1, u_2 \in U$, $v_1, v_2 \in V$ and $\lambda, \mu \in K$.

A bilinear form $< \ , \ >$ on $V \times V$ is called a *bilinear form on V*. A bilinear form $< \ , \ >$ on V is said to be *symmetric* if $<x,y> = <y,x>$ for all $x, y \in V$.

If $< \ , \ >$ is a bilinear form on $U \times V$ which satisfies

$$<u, v'> = 0 \text{ for all } u \in U \text{ implies } v' = 0, \text{ and}$$

$$<u', v> = 0 \text{ for all } v \in V \text{ implies } u' = 0,$$

then $< \ , \ >$ is called a *nondegenerate* bilinear form.

Let U and V be finite dimensional K-spaces. Given a basis $\{u_1, ..., u_m\}$ of U and a basis $\{v_1, .., v_n\}$ of V, if $< \ , \ >$ is any bilinear form on $U \times V$, then there is an associated $m \times n$ matrix $A = (a_{ij})$ with

$$a_{ij} = <u_i, v_j>, \ i = 1, ..., m, \ j = 1, ..., n,$$

and $< \ , \ >$ is completely determined by A, that is, given

(1) $$u = \sum_{i=1}^{m} \lambda_i u_i, \quad v = \sum_{j=1}^{n} \mu_j v_j,$$

it follows from the bilinear property of $<\ ,\ >$ that

$$<u,v> = <\sum \lambda_i u_i, \sum \mu_j v_j>$$

$$= \sum \lambda_i \mu_j <u_i, v_j>$$

$$= \sum \lambda_i a_{ij} \mu_j$$

$$= (\lambda_1, ..., \lambda_m) \begin{pmatrix} a_{11} & \cdots & a_{1n} \\ \vdots & \vdots & \vdots \\ a_{m1} & \cdots & a_{mn} \end{pmatrix} \begin{pmatrix} \mu_1 \\ \vdots \\ \mu_n \end{pmatrix}.$$

Conversely, any $m \times n$ matrix over K yields a bilinear form on $U \times V$ in this way.

Now, suppose that $\{u'_1, ..., u'_m\}$ and $\{v'_1, ..., v'_n\}$ are new bases for U and V respectively, and that

(2) $$u = \sum_{i=1}^{m} \lambda'_i u'_i, \quad v = \sum_{j=1}^{n} \mu'_j v'_j.$$

Then it follows from a change of bases and (1) that

$$(\lambda_1, ..., \lambda_m) = (\lambda'_1, ..., \lambda'_m)P,$$

$$(\mu_1, ..., \mu_n) = (\mu'_1, ..., \mu'_n)Q$$

for some invertible matrices $P = P_{m \times m}$, $Q = Q_{n \times n}$. Thus,

$$<u,v> = (\lambda'_1, ..., \lambda'_m) P A Q^t \begin{pmatrix} \mu'_1 \\ \vdots \\ \mu'_n \end{pmatrix},$$

and consequently, the matrix referred to the new bases is PAQ^t.

5.7. Theorem (i) Let U and V be finite dimensional vector spaces over a field K, where $\dim_K U = m$ and $\dim_K V = n$. If $<\ ,\ >$ is any nondegenerate bilinear form on $U \times V$, then $m = n$, and for any basis $\{u_1, ..., u_n\}$ of U there exists a unique basis $\{v_1, ..., v_n\}$ of V such that

$$<u_i, v_j> = \delta_{ij} = \begin{cases} 0, \text{if } i \neq j, \\ 1, \text{if } i = j. \end{cases}$$

(ii) A bilinear form $<\ ,\ >$ on a finite n-dimensional K-space V is nondegenerate if and only if the associated matrix $A = (a_{ij})$, where $a_{ij} =< v_i, v_j >$, is invertible for any basis $\{v_1, ..., v_n\}$ of V.

Proof (i) Let $\{u_1, ..., u_n\}$ be any basis of U. Consider the linear mapping induced by $<\ ,\ >$

$$\sigma: V \longrightarrow K^m = \{(\lambda_1, ..., \lambda_m) \mid \lambda_i \in K\}$$

$$v \mapsto (< u_1, v >, ..., < u_m, v >)$$

Then σ is injective because $<\ ,\ >$ is nondegenerate. Thus, $n = \dim_K V \leq \dim_K U = m$. Similarly we also have $m \leq n$. Hence $m = n$.

Note that σ is now an isomorphism. If we use the standard basis $\{e_1, ..., e_m\}$ of K^m, where

$$e_j = (\underbrace{0, ..., 0}_{j-1}, 1, 0, ..., 0),\ j = 1, ..., m,$$

and write v_j for the inverse image of e_j under σ, then it is clear that $\{v_1, ..., v_m\}$ is a basis for V and $< u_i, v_j >= \delta_{ij}$, $i, j = 1, ..., m$. If $\{v'_1, ..., v'_m\}$ is another basis of V with this property, then $\sigma(v_i - v'_i) = 0$ implies $v_i = v'_i$, $i = 1, ..., m$, because σ is injective.

(ii) This follows from part (i) and previous discussion on the associated matrices of $<\ ,\ >$ with respect to given bases.

Exercises

1. Complete the proof of Proposition 5.4.
2. Let $d \in \mathbb{Z}$ be square-free, $K = \mathbb{Q}(\sqrt{d})$. Find all $\mathbb{Q}$-linear ring monomorphisms $K \to \mathbb{C}$.
3. Let $K = \mathbb{Q}(\sqrt{d})$ be as in exercise 2 above, $\alpha = r + s\sqrt{d}$. Show that $T_{K/\mathbb{Q}}(\alpha) = 2r$ and $N_{K/\mathbb{Q}}(\alpha) = r^2 - s^2 d$. (Compare with section 3, exercise 6.)
4. Let $\vartheta = \sqrt{2} + \sqrt{3}$, $K = \mathbb{Q}(\vartheta)$. Find the minimal polynomial $p(x)$ of ϑ. (Hint: Note that $\mathbb{Q}(\sqrt{2}), \mathbb{Q}(\sqrt{3}) \subset K$. Any $\mathbb{Q}$-linear ring monomorphism $\sigma: K \to \mathbb{C}$ induces a $\mathbb{Q}$-linear ring monomorphism $\sigma_1: \mathbb{Q}(\sqrt{2}) \to \mathbb{C}$ and a $\mathbb{Q}$-linear ring monomorphism $\sigma_2: \mathbb{Q}(\sqrt{3}) \to \mathbb{C}$, such that $\sigma(\vartheta) = \sigma_1(\sqrt{2}) + \sigma_2(\sqrt{3})$. The answer is $p(x) = x^4 - 10x^2 + 1$.)

Can you generalize this result to $\vartheta = \sqrt{p} + \sqrt{q}$ for arbitrary square-free $p \neq q$?

6. Free Abelian Groups of Finite Rank

Let G be an abelian group with the binary additive operation $+$ and the identity element 0. For $g \in G$, we write $\mathbb{Z}g$ for the cyclic subgroup of G generated by g, and consequently, we write $\sum_{g \in \Omega} \mathbb{Z}g$ for the subgroup of G generated by a nonempty subset $\Omega \subseteq G$.

A subset $\Omega = \{g_i\}_{i \in J}$ of G is said to be $\mathbb{Z}$-*linearly independent* if for any finitely many $g_{i_1}, g_{i_2}, ..., g_{i_n} \in \Omega$, there do not exist $s_1, ..., s_n \in \mathbb{Z}$, not all zero, such that $s_1 g_{i_1} + s_2 g_{i_2} + \cdots + s_n g_{i_n} = 0$. If Ω is not $\mathbb{Z}$-linearly independent, then it is $\mathbb{Z}$-*linearly dependent*.

6.1. Definition Let $\Omega = \{g_i\}_{i \in J} \subset G$. If $G = \sum_{g_i \in \Omega} \mathbb{Z}g_i$ and Ω is $\mathbb{Z}$-linearly independent, then G is called a *free abelian group* and Ω is called a $\mathbb{Z}$-*basis* of G, or just a *basis* of G.

Below we focus on free abelian groups with finite $\mathbb{Z}$-basis.

6.2. Proposition If a free abelian group G has two bases $\{g_1, ..., g_n\}$ and $\{h_1, ..., h_m\}$, then $m = n$.

Proof Suppose $m < n$. Then, as dealing with vector bases over a field in linear algebra, after expressing each g_i as a $\mathbb{Z}$-linear combination of h_i's, we may derive, by passing to $\mathbb{Q}$, that $\{g_1, ..., g_n\}$ is $\mathbb{Z}$-linearly dependent, a contradiction. Hence, $m \geq n$. By symmetry, $n \geq m$. Thus, $m = n$ as desired. $\square$

6.3. Definition An abelian group G with a basis of n elements is called a *free abelian group of* $\mathbb{Z}$-*rank* n, or just a *free abelian group of rank* n.

$\mathbb{Z}$ is a free abelian group of rank 1 and $\{1\}$ is a basis for $\mathbb{Z}$. The direct sum

$$\mathbb{Z}^n = \mathbb{Z} \oplus \mathbb{Z} \oplus \cdots \oplus \mathbb{Z} = \left\{ (k_1, ..., k_n) \mid k_i \in \mathbb{Z} \right\}$$

of n copies of $\mathbb{Z}$, where

$$(k_1, ..., k_n) + (\ell_1, ..., \ell_n) = (k_1 + \ell_1, ..., k_n + \ell_n),$$

$$m(k_1, ..., k_n) = (mk_1, ..., mk_n), \ m \in \mathbb{Z},$$

is a free abelian group of rank n with the standard basis

$$\left\{ e_i = (\underbrace{0,...,0}_{i-1},1,0,...,0) \;\middle|\; i = 1,...,n \right\}.$$

6.4. Proposition Any finitely generated abelian group $G = \sum_{i=1}^{n} \mathbb{Z}g_i$ is a homomorphic image of some free abelian group of rank n. If G is free and $\{g_1,...,g_n\}$ is a basis of G, then $G \cong \mathbb{Z}^n$.

Proof Exercise. □

In view of Proposition 6.4, from now on we write

$$G = \bigoplus_{i=1}^{n} \mathbb{Z}g_i$$

for the free abelian group G with basis $\{g_1,...,g_n\}$.

Let $M_n(\mathbb{Z})$ be the set of all $n \times n$ matrices over $\mathbb{Z}$. If $A \in M_n(\mathbb{Z})$ and $\det(A) = \pm 1$, then we say that A is *unimodular*.

If $A \in M_n(\mathbb{Z})$ is unimodular, then A is invertible and

$$A^{-1} = \frac{1}{\det(A)} A^* = \pm A^*$$

where A^* is the adjoint matrix of A. Clearly, the construction of A^* implies $A^* \in M_n(\mathbb{Z})$. It follows that $A^{-1} \in M_n(\mathbb{Z})$.

6.5. Lemma If $\{u_1,...,u_n\}$ is a basis for the free abelian group G, then $\{v_1,...,v_n\}$, where

$$v_i = \sum_{j}^{n} a_{ij} u_j, \; a_{ij} \in \mathbb{Z}, \; i = 1,...,n,$$

is a basis for G if and only if $A = (a_{ij})$ is unimodular.

Proof With the help of the above remark, this is an easy exercise. □

6.6. Theorem Let G be a free abelian group of rank n and H a nonzero subgroup of G. Then the following hold:
(i) H is free of rank $s \leq n$.

(ii) There exist a basis $\{g_1, ..., g_n\}$ for G and integers $\ell_1, ..., \ell_s \in \mathbb{Z}^+$ such that $\{\ell_1 g_1, ..., \ell_s g_s\}$ is a basis for H.

Proof We prove the theorem by induction on the rank of G. If rank$G = 1$, then $G \cong \mathbb{Z}$ and the conclusions (i) and (ii) are clear.

Suppose the assertions (i) and (ii) are true for any free abelian group of rank $< n$.

Let G be a free abelian group of rank n and let H be a nonzero subgroup of G. Then, with respect to a fixed basis $\{e_1, ..., e_n\}$ of G, H contains elements

(1) $$h = k_1 e_1 + \cdots + k_n e_n, \text{ some } k_i\text{'s are positive.}$$

Choose a basis $\{x_1, ..., x_n\}$ of G such that ℓ_1 is the smallest positive coefficient with respect to (1), and rearrange the members of this basis (if necessary) so that H contains an element of the form

$$f_1 = \ell_1 x_1 + m_2 x_2 + \cdots + m_n x_n.$$

On division by ℓ_1, write

(2) $$m_i = \ell_1 q_i + r_i, \quad q_i, r_i \in \mathbb{Z}, \ 0 \leq r_i < \ell_1, \ 2 \leq i \leq n.$$

If we define

$$g_1 = x_1 + q_2 x_2 + \cdots + q_n x_n$$

$$g_2 = x_2$$

$$\vdots$$

$$g_n = x_n,$$

then

$$\begin{pmatrix} g_1 \\ g_2 \\ \vdots \\ g_n \end{pmatrix} = \begin{pmatrix} 1 & q_2 & \cdots & q_n \\ 0 & 1 & \cdots & 0 \\ \vdots & \vdots & \cdots & \vdots \\ 0 & 0 & \cdots & 1 \end{pmatrix} \begin{pmatrix} x_1 \\ x_2 \\ \vdots \\ x_n \end{pmatrix}$$

where the square matrix is clearly unimodular. By Lemma 6.5, $\{g_1, x_2, ..., x_n\}$

is a basis of G. With respect to this new basis and previous formula (2),

$$f_1 = \ell_1 x_1 + m_2 x_2 + \cdots + m_n x_n$$

$$= \ell_1(g_1 - q_2 x_2 - \cdots - q_n x_n) + m_2 x_2 + \cdots + m_n x_n$$

$$= \ell_1 g_1 + (m_2 - \ell_1 q_2) x_2 + \cdots + (m_n - \ell_1 q_n) x_n$$

$$= \ell_1 g_1 + r_2 x_2 + \cdots + r_n x_n.$$

By the choice of ℓ_1, we must have $r_2 = \cdots = r_n = 0$. It follows that

(3) $$f_1 = \ell_1 g_1.$$

Now, with respect to the new basis $\{g_1, x_2, ..., x_n\}$, each $h \in H$ has the expression

$$h = c_1 g_1 + c_2 x_2 + \cdots + c_n x_n, \ c_i \in \mathbb{Z}.$$

Write $c_1 = \ell_1 q + r$ with $q, r \in \mathbb{Z}$, $0 \le r < \ell_1$. Then by the above (3), H contains

$$h - q f_1 = \ell_1 q g_1 + r g_1 - \ell_1 q g_1 + c_2 x_2 + \cdots + c_n x_n$$

$$= r g_1 + c_2 x_2 + \cdots + c_n x_n.$$

Again by the choice of ℓ_1 we have $r = 0$. This yields

$$q f_1 + c_2 x_2 + \cdots + c_n x_n = h \in \mathbb{Z} f_1 + G^* \text{ with } G^* = \bigoplus_{j=2}^{n} \mathbb{Z} x_j.$$

Thus

$$\varphi: H \longrightarrow G^*$$

$$h \mapsto c_2 x_2 + \cdots + c_n x_n$$

defines a group homomorphism and $H \cong \mathbb{Z} f_1 + \varphi(H)$. Note that $\varphi(H)$ is a subgroup of G^* that is free of rank $n - 1$. By the induction hypothesis, there exist basis $\{g_2, ..., g_n\}$ of G^* and integers $\ell_2, ..., \ell_s \in \mathbb{Z}^+$, where $s \le n - 1$, such that $\{\ell_2 g_2, ..., \ell_n g_s\}$ forms a basis for $\varphi(H)$. Thus, $\{\ell_1 g_1 = f_1, \ell_2 g_2, ..., \ell_s g_s\}$ forms a basis for H, as desired. $\square$

6.7. Theorem Let G be a free abelian group of rank n and H a subgroup of G. The following statements hold:

(i) G/H is finite if and only if $\mathrm{rank} G = \mathrm{rank} H$.
(ii) If $\mathrm{rank} G = \mathrm{rank} H = n$, $\{x_1,...,x_n\}$ is a basis for G, $\{y_1,...,y_n\}$ is a basis for H, and

$$y_i = \sum_{j=1}^n a_{ij} x_j, \ a_{ij} \in \mathbb{Z}, \ i = 1,...,n,$$

then the number of elements of G/H is equal to $|\det(A)|$, where $A = (a_{ij})$.

Proof (i) By Theorem 6.6, choose a basis $\{g_1,...,g_n\}$ of G and a basis $\{f_1,...,f_s\}$ of H with $f_i = \ell_i g_i$ and $\ell_i \in \mathbb{Z}^+$, $i = 1,...,s \leq n$. Thus

$$G = \mathbb{Z} g_1 \oplus \mathbb{Z} g_2 \oplus \cdots \oplus \mathbb{Z} g_n$$

$$H = \mathbb{Z}\ell_1 g_1 \oplus \mathbb{Z}\ell_2 g_2 \oplus \cdots \oplus \mathbb{Z}\ell_s g_s$$

and we have the group isomorphism

$$G/H \xrightarrow{\cong} \mathbb{Z}_{\ell_1} \oplus \mathbb{Z}_{\ell_2} \oplus \cdots \oplus \mathbb{Z}_{\ell_s} \oplus (\mathbb{Z} g_{s+1} \oplus \cdots \oplus \mathbb{Z} g_n)$$

$$\overline{\sum_{i=1}^s k_i g_i + \sum_{j=s+1}^n k_j g_j} \mapsto \sum_{i=1}^s \overline{k_i} + \sum_{j=s+1}^n k_j g_j$$

It follows that G/H is finite if and only if $n = s$.
(ii) By the proof of part (i), if G/H is finite then it has exactly $\ell_1 \ell_2 \cdots \ell_n$ elements. Employing the chosen bases in part (i), we have

$$g_i = \sum_{j=1}^n b_{ij} x_j$$

$$f_i = \sum_{j=1}^n c_{ij} g_j = \ell_i g_i$$

$$y_i = \sum_{j=1}^n d_{ij} f_j$$

Then $(b_{ij}) = B$ and $(d_{ij}) = D$ are unimodular, and

$$C = (c_{ij}) = \begin{pmatrix} \ell_1 & 0 & \cdots & 0 \\ 0 & \ell_2 & \cdots & 0 \\ \vdots & \vdots & \cdots & \vdots \\ 0 & 0 & \cdots & \ell_n \end{pmatrix}.$$

Taking the matrix $A = (a_{ij})$ from the assumption of part (ii) into account, we get $A = BCD$ and $\det(A) = \det(B)\det(C)\det(D)$. Therefore, $|\det(A)| = \ell_1 \ell_2 \cdots \ell_n$.

Exercises

1. Let $i = \sqrt{-1}$. Show that $\mathbb{Z}[i] = \{a + bi \mid a, b \in \mathbb{Z}\} \cong \mathbb{Z} \oplus \mathbb{Z}$.
2. Complete the proof of Proposition 6.4.
3. Complete the proof of Lemma 6.5.
4. An abelian group G is said to be *torsion-free* if G does not have finite order nonzero element. Show that a finitely generated torsion-free abelian group is free of finite rank. (Hint: Use Proposition 6.4 and refer to the proof of Theorem 6.7.)
5. Show that a finitely generated abelian group G is either finite or isomorphic to the direct sum of a free abelian group of finite rank and a finite abelian group.

7. Noetherian Modules

Let R be a ring.

7.1. Definition Let M be an abelian group with the binary additive operation $+$ and the identity element 0. We say that M is an *R-module* if there is a mapping

$$\alpha : R \times M \longrightarrow M$$

$$(r, m) \mapsto \alpha(r, m) = rm$$

(called the *R-action* on M) satisfying
(M1) $(r + s)m = rm + sm$,
(M2) $r(m + m') = rm + rm'$,
(M3) $r(sm) = (rs)m$,

(M4) $1m = m$
for all $r, s \in R$ and $m, m' \in M$.

By definition, a $\mathbb{Z}$-module is nothing but an abelian group M (binary operation is written additively). Conversely, given an (additive) abelian grooup M, M can be made into a $\mathbb{Z}$-module by defining

$$0m = 0 \text{ and } 1m = m \text{ for } m \in M,$$

then inductively

$$(n+1)m = nm + m \text{ for } n \in \mathbb{Z}^+, \ m \in M,$$

and

$$(-n)m = -nm \text{ for } n \in \mathbb{Z}^+, \ m \in M.$$

If $R = K$ is a field, then an R-module is nothing but a K-vector space. In this sense we may view an R-module as the generalization of a vector space. However, since not every nonzero element in an arbitrary ring R is a unit, many of the techniques developed in vector space theory cannot be performed directly to deal with R-modules.

From the definition it is clear that if M is an R-module then every $r \in R$ defines an endomorphism of the abelian group M, that is, $\rho_r \colon M \to M$ with $\rho_r(m) = rm$. One easily checks that this yields a ring homomorphism $\sigma \colon R \to \operatorname{End}_{\mathbb{Z}} M$ with $\sigma(r) = \rho_r$, where $\operatorname{End}_{\mathbb{Z}} M$ is the ring of endomorphisms of M. Conversely, if M is an abelian group then any ring homomorphism $\varphi \colon R \to \operatorname{End}_{\mathbb{Z}} M$ induces an R-module structure: $rm = \varphi(r)(m)$. This is the idea of modern representation theory of rings and algebras.

Two special kinds of module will be used frequently in the follow-up chapters:

- If R is a subring of a ring S (note that $1_R = 1_S$ by our convention made on rings), then S is an R-module with the action given by the ring multiplication.
- If I is an ideal of the ring R. Then I is an R-module with the action given by the ring multiplication.

Let M be an R-module and N an (additive) subgroup of M. If $rx \in N$ for all $r \in R$ and $x \in N$, then N is called an R-*submodule* of M.

Given a family $\{N_i\}_{i \in I}$ of R-submodules of M, the sum $\sum_{i \in I} N_i$ of subgroups forms an R-submodule in a natural way; and the intersection

$\bigcap_{i \in I} N_i$ is an R-submodule.

Given an R-submodule of a module M, an R-action on the quotient group M/N is defined as

$$r\overline{m} = \overline{rm}, \ r \in R, \quad \overline{m} \in M/N.$$

With the R-action defined above, M/N is called the *quotient R-module determined by N*.

Let M and N be R-modules. An *R-module homomorphism* from M to N is a homomorphism of abelian groups $\psi \colon M \to N$ satisfying $\psi(rm) = r\psi(m)$ for all $r \in R$ and $m \in M$. It can be verified directly that $\mathrm{Ker}\psi$ is an R-submodule of M, that $\mathrm{Im}\psi$ is a submodule of N, and furthermore, that the following R-module isomorphism theorems hold:

(a) $M/\mathrm{Ker}\psi \cong \mathrm{Im}\psi$.
(b) Let A, B be submodules of the R-module M. Then, $(A + B)/B \cong A/(A \cap B)$.
(c) Let A, B be submodules of the R-module M. If $A \subseteq B$ then $(M/A)/(B/A) \cong M/B$.
(d) Let N be a submodule of the R-module M. Then there is a bijection between the submodules of M which contain N and the submodules of M/N:

$$\alpha \colon \quad A \longleftrightarrow (A + N)/N$$

such that $\alpha(A + B) = \alpha(A) + \alpha(B)$ and $\alpha(A \cap B) = \alpha(A) \cap \alpha(B)$ for all submodules A, B of M containing N.

Let $S \subseteq M$, where M is an R-module, and $T \subseteq R$. Put

$$TS = \left\{ \text{finite sums } \sum r_i m_i \ \middle| \ r_i \in T, \ m_i \in S \right\}.$$

With notation as above, the reader is also asked to check the following statements.

(e) If $T = R$, then RS forms an R-submodule of M; moreover, $RS = \sum_{m_i \in S} Rm_i$ and it is the smallest R-submodule of M containing S.
(f) If T is an ideal of R, then TM forms an R-submodule.

The R-submodule $N = RS$ obtained in part (e) above is called an *R-submodule of M generated by S*, where S is called a *set of generators* of N. If $M = RS$ with a finite set of generators $S = \{m_1, ..., m_s\}$, then $M = \sum_{i=1}^{s} Rm_i$ and is called a *finitely generated R-module*.

Given a family of R-modules $\{M_i\}_{i \in J}$, the direct sum of abelian groups

$$\bigoplus_{i \in J} M_i = \left\{ (m_i)_{i \in J} \mid 0 \neq m_i \in M_i \text{ for only finitely many } m_i \right\}$$

is an R-module, where

$$(m_i)_{i \in J} + (m'_i)_{i \in J} = (m_i + m'_i)_{i \in J},$$

$$r(m_i)_{i \in J} = (rm_i)_{i \in J}, \ r \in R,$$

and is called the *direct sum* of $\{M_i\}_{i \in J}$.

For the direct sum $\oplus_{i \in J} M_i$ of given R-modules defined above, it is not hard to see that there is an injective R-module homomorphism $M_i \to \oplus_{i \in J} M_i$ with $m_i \mapsto (x_i)_{i \in J}$, where $x_i = m_i$ and $x_j = 0$ for $j \neq i$. Hence M_i is isomorphic to a submodule of $\oplus_{i \in J} M_i$. Conversely, let $\{N_i\}_{i \in J}$ be a family of submodules of some R-module M, and let $N = \sum_{i \in J} N_i$. Then

$$\phi: \bigoplus_{i \in J} N_i \longrightarrow N = \sum_{i \in J} N_i$$

$$(x_i)_{i \in J} \mapsto \sum x_i$$

defines an R-module homomorphism. If ϕ is an isomorphism then N is said to be the *direct sum of its submodules* N_i, $i \in J$, and we also write $N = \oplus_{i \in J} N_i$.

Let M be an R-module and suppose that $M = \oplus_{i \in J} M_i$ for some submodules $M_i \subset M$, $i \in J$. Then it is clear that every element $m \in M$ has a *unique* expression $m = \sum m_i$, i.e., $\sum m_i = 0$ if and only if $m_i = 0$.

7.2. Definition An R-module M is said to be *free* if there are $\xi_i \in M$, $i \in J$, such that $M = \oplus_{i \in J} R\xi_i$, where $\{\xi_i\}_{i \in J}$ is called an *R-basis* of M.

Example (i) Any vector space V over a field K is a free K-module. Any free abelian group (as defined in section 6) is a free $\mathbb{Z}$-module.

(ii) For any set J of indices, $F = \oplus_{i \in J} R_i$ with $R_i \cong R$ (as R-modules) is a free R-module.

7.3. Proposition Any R-module M is the homomorphic image of some free R-module.

Proof The R-module homomorphism $\oplus_{m\in M} R_m \to M = \sum_{m\in M} Rm$ defined by $\sum r_m \mapsto \sum r_m m$ does the job. $\square$

An R-module M is said to be *Noetherian* if every R-submodule of M is finitely generated.

7.4. Theorem For an R-module M, the following statements are equivalent.
(i) M is Noetherian.
(ii) For any ascending chain
$$M_1 \subseteq M_2 \subseteq \cdots \subseteq M_n \subseteq \cdots$$
of R-submodules in M, there is some k such that $M_k = M_j$ for all $j \geq k$.
(iii) Every nonempty set of R-submodules has a maximal element with respect to $\subseteq$.

Proof Exercise (see the proof of Theorem 1.1). $\square$

7.5. Theorem (i) Let $\varphi\colon M \to H$ be an onto R-module homomorphism. If R is Noetherian then so are $\operatorname{Ker}\varphi$ and H.
(ii) Let N be an R-submodule of the R-module M. Then M is Noetherian if and only if N and M/N are Noetherian.

Proof To better understand the argumentation, the reader is reminded to bear the foregoing R-isomorphism theorems (a)–(d) in mind.
(i) This follows from the fact that ascending chains of R-submodules in $\operatorname{Ker}\varphi$ and H correspond to some ascending chains of submodules in M.
(ii) If M is Noetherian then so are N and M/N by part (i). Now let
$$M_1 \subset M_2 \subset \cdots \subset M_n \subset \cdots$$
be an ascending chain of R-submodules of M. Then
$$N \cap M_1 \subset N \cap M_2 \subset \cdots \subset N \cap M_n \subset \cdots$$
is a chain of R-submodules in N and for some $\ell \geq 1$

(1) $\quad N \cap M_\ell = N \cap M_{\ell+i},\ i = 1, 2, \ldots.$

On the other hand, we also have a chain of R-submodules in M/N
$$\frac{M_1 + N}{N} \subset \frac{M_2 + N}{N} \subset \cdots \subset \frac{M_n + N}{N} \subset \cdots$$

and (without loss of generality) for $\ell \geq 1$

(2) $$\frac{M_\ell + N}{N} = \frac{M_{\ell+i} + N}{N}, \ i = 1, 2, \ldots$$

Thus, for $m \in M_{\ell+i}$, formula (2) implies $m = m' + x$ for some $m' \in M_\ell$ and $x \in N$. But then

$$x = m - m' \subset M_{\ell+i} \cap N = M_\ell \cap N$$

by (1) above. It follows that $m - m' = m''$ with $m'' \in M_\ell \cap N$, and consequently $m = m' + m'' \in M_\ell$. This shows that $M_j = M_\ell$ for $j \geq \ell$, that is, M is Noetherian. □

7.6. Theorem (i) Given finitely many Noetherian R-modules $M_1, \ldots, M_s$, the direct sum $\oplus_{i=1}^s M_i$ is a Noetherian R-module.
(ii) If R is a Noetherian ring and M is a finitely generated R-module, then every submodule of M is Noetherian, in particular, M is Noetherian.

Proof (i) Set $M = M_1 \oplus M_2$. Then M_1 and $M/M_1 \cong M_2$ are Noetherian by the assumption. It follows from Proposition 7.5(ii) that M is Noetherian. Now an induction on s shows that $\oplus_{i=1}^s M_i$ is Noetherian.
(ii) Suppose $M = \sum_{i=1}^s R\xi_i$, $\xi_i \in M$. Then there is an onto R-module homomorphism

$$\underbrace{R \bigoplus R \bigoplus \cdots \bigoplus R}_{s} \longrightarrow M = \sum_{i=1}^s R\xi_i$$

$$\sum_{i=1}^s r_i \quad \mapsto \quad \sum_{i=1}^s r_i \xi_i$$

So the conclusion now follows from part (i) and Theorem 7.5. □

We complete this chapter with the celebrated Krull's intersection theorem.

7.7. Theorem (Krull) Let R be a Noetherian ring and I an ideal of R. Given a finitely generated R-module M, let

$$U = \bigcap_{n=1}^\infty I^n M.$$

Then $IU = U$.

Proof First note that every $I^n M$, hence U and IU are R-submodules. So it is clear that we need only to show $U \subseteq IU$. For this purpose, noticing $U \cap IU = IU$, let us consider

$$\Omega = \Big\{ S \;\Big|\; S \text{ a submodule of } M, S \cap U = IU \Big\}.$$

Then Ω has a maximal member, say S, with respect to $\subseteq$ on submodules, for M is Noetherian by Theorem 7.6.

Claim For the maximal S obtained above, there is some n such that $I^n M \subseteq S$, and consequently, $U = I^n M \cap U \subseteq S \cap U = IU$.

To find the above claimed n, let $I = \sum_{i=1}^{s} R\xi_i$, $\xi_i \in I$. If we can find, for each ξ_i, some n_i such that

(∗) $\qquad\qquad\qquad \xi_i^{n_i} M \subseteq S,$

then there will be some n, large enough, such that $I^n M \subseteq S$. As a matter of fact, we may reach the above mentioned property (∗) for any $a \in I$. To see this, define, for each $k \geq 1$, the R-submodule

$$M_k = \Big\{ m \in M \;\Big|\; a^k m \in S \Big\}.$$

Then we obtain an ascending chain

$$M_1 \subseteq M_2 \subseteq \cdots \subseteq M_q \subseteq \cdots$$

and there is some z such that $M_z = M_j$ for all $j \geq z$. For this fixed z, obviously $IU \subseteq (a^z M + S) \cap U$. On the other hand, if $u \in (a^z M + S) \cap U$, then $u = a^z m + v$ with $m \in M$ and $v \in S$. Hence $au \in aU \subseteq IU \subseteq S$, and $a^{z+1} m \in S$. This shows that $m \in M_{z+1} = M_z$, and it follows that $a^z m \in S$. But this yields $u \in S \cap U = IU$, and consequently $IU = (a^z M + S) \cap U$. If $a^z M + S = M$, then $U = IU$; otherwise, by the maximality of S in Ω, $a^z M \subseteq S$, as desired. $\square$

7.8. Corollary Let R be a Noetherian domain, and let M be a finitely generated torsion-free R-module, i.e., for $r \in R$ and $m \in M$, $rm = 0$ implies $r = 0$ or $m = 0$. Then

$$\bigcap_{n=1}^{\infty} I^n M = \{0\}.$$

Proof This follows from Theorem 7.7 and later exercise 6.

Exercises

1. Find all $\mathbb{Z}$-submodules of $\mathbb{Z}$ and all $\mathbb{Z}$-module homomorphisms $\mathbb{Z} \to \mathbb{Z}$.
2. Complete the proof of Proposition 7.4.
3. Let R be a domain with the field of fractions K. Let $\lambda \in R$ be nonzero and nonunit. Show that $R[\frac{1}{\lambda}]$, the subring of K generated by $\frac{1}{\lambda}$ over R, is not a finitely generated R-module. (Hint: If there was a finite set of generators, then $1, \frac{1}{\lambda}, \frac{1}{\lambda^2}, ..., \frac{1}{\lambda^s}$ would be a set of generators for some $s > 0$. After expressing $\frac{1}{\lambda^{s+1}}$ as an R-linear combination of the foregoing generators, see what happens.)
4. Show that there is no nonzero $\mathbb{Z}$-module homomorphism $\mathbb{Q} \to \mathbb{Z}$.
5. Let R be a ring and let $I_1, ..., I_s$ be finitely many ideals of R. Suppose that R/I_j is Noetherian, $j = 1, ..., s$, and that $\cap_{j=1}^s I_j = \{0\}$. Show that R is Noetherian. (Hint: Consider the R-module homomorphism $R \to \oplus_{j=1}^s (R/I_j)$ with $r \mapsto \sum x_j$, where $x_j = \overline{r} \in R/I_j$, $j = 1, ..., s$.)
6. For any ring R, one may also define matrices $(r_{ij})_{m \times n}$ of finite order with entries $r_{ij} \in R$, define addition and multiplication of matrices, and define the determinant, adjoint and inverse of a square matrix, as in classical linear algebra.

    Let $M = \sum_{i=1}^s R\xi_i$ be a finitely generated R-module, where $\xi_i \in M$, $i = 1, ..., s$, and let I be an ideal of R. Show that if $IM = M$ then there is some $r \in R$ such that $rM = \{0\}$ and $1 - r \in I$. (Hint: Note that $IM = M$ implies $\xi_i = \sum_{j=1}^s a_{ij}\xi_j$, $i = 1, ..., s$, $a_{ij} \in I$. Thus

$$\begin{pmatrix} a_{11} - 1 & a_{12} & \cdots & a_{1s} \\ a_{21} & a_{22} - 1 & \cdots & a_{2s} \\ \vdots & & \vdots & \\ a_{s1} & a_{s2} & \cdots & a_{ss} - 1 \end{pmatrix} \begin{pmatrix} \xi_1 \\ \xi_2 \\ \vdots \\ \xi_s \end{pmatrix} = \begin{pmatrix} 0 \\ 0 \\ \vdots \\ 0 \end{pmatrix}.$$

Multiplying by the adjoint $(a_{ij})^*$ of (a_{ij}), it follows that $\det(a_{ij})M = \{0\}$, where $\det(a_{ij}) = 1 - a$ for some $a \in I$.)

7. Let R and K be as in exercise 3 above. Use problem 6 to show that K is not a finitely generated R-module. (Hint: Take a nonzero nonunit $\lambda \in R$ and note that $\lambda K = K$.)
8. Let $R = A[x_1, ..., x_n]$ be the polynomial ring in $x_1, ..., x_n$ over a ring A. Let $R_i = \sum_{\alpha_1 + \cdots + \alpha_n = i} A x_1^{\alpha_1} \cdots x_n^{\alpha_n}$, $i \in \mathbb{N}$, which is called the ith *homogeneous part* of R. Show that, as A-modules, $R = \oplus_{i \in \mathbb{N}} R_i$, and that, as subsets, $R_i R_j = R_{i+j}$, $i, j \in \mathbb{N}$.

Chapter 2
Local Rings, DVRs, and Localization

Commutative algebra studies various problems related to "factorizations into irreducible components" and "zeros of polynomials" by means of rings and their modules. In this chapter and the next, we introduce the fundamental structures and methods that are essential in demonstrating the principles of commutative algebra such as "singularities versus normalization" and "global concern versus local solutions".

1. Spec$R$, m-Spec$R$, and Radicals

Let R be a ring.

1.1. Definition (i) An ideal $P \subsetneq R$ is called a *prime ideal* if P has the property:

$$\text{for } a, b \in R, \ ab \in P \text{ implies } a \in P \text{ or } b \in P.$$

Write Spec$R$ for the set of all prime ideals of R and call it the *prime spectrum* of R.
(ii) An ideal $M \subsetneq R$ is said to be *maximal* if for any ideal I of R,

$$M \subseteq I \text{ implies } M = I \text{ or } I = R.$$

Write m-Spec$R$ for the set of all maximal ideals of R and call it the *maximal spectrum* of R.

A subset $S \subset R$ is called a *multiplicative set* if $1 \in S$ and $a, b \in S$ implies $ab \in S$.

1.2. Proposition (i) An ideal P of R is prime if and only if R/P is a domain, or equivalently, if and only if $S = R - P$ is a multiplicative set.
(ii) An ideal M of R is maximal if and only if R/M is a field. Consequently, R is a field if and only if $\langle 0 \rangle$ is maximal in R.

Proof By definition, both parts (i) and (ii) may be verified directly. □

Example 1 (i) If $\varphi\ R \to B$ is a *nonzero* ring homomorphism from R to a domain B, then $\mathrm{Ker}\varphi = P$ is a prime ideal of R by Proposition 1.2(ii).

(ii) Let R be a domain. Then $\{0\}$ is a prime ideal of R. If $0 \neq a \in R$, then the principal ideal $\langle a \rangle$ is a prime ideal if and only if a is a prime. Thus, a PID R has $\mathrm{Spec}R = \{\langle 0 \rangle\} \cup \{\langle p \rangle \mid p \in R$ a prime$\}$. In particular, $\mathrm{Spec}\mathbb{Z} = \{\langle 0 \rangle\} \cup \{\langle p \rangle \mid p \in \mathbb{Z}$ a prime number$\}$; and if $K[x]$ is the polynomial ring in x over a field K, then $\mathrm{Spec}K[x] = \{\langle 0 \rangle\} \cup \{\langle f \rangle \mid f$ is irreducible in $K[x]\}$ (in case K is algebraically closed, $\mathrm{Spec}K[x] = \{\langle x - \lambda \rangle \mid \lambda \in K\} \cup \{\langle 0 \rangle\}$).

(iii) In the polynomial ring $K[x_1, ..., x_n]$ in $x_1, ..., x_n$ over a field K,

$$\langle 0 \rangle \subset \langle x_1 \rangle \subset \langle x_1, x_2 \rangle \subset \langle x_1, x_2, x_3 \rangle \subset \cdots \subset \langle x_1, ..., x_n \rangle$$

is a chain of prime ideals.

For any n-tuple $P = (a_1, ..., a_n) \in K^n$, define the function

$$\psi_P : K[x_1, ..., x_n] \longrightarrow K$$

$$f(x_1, ..., x_n) \mapsto f(P) = f(a_1, ..., a_n)$$

where if $f = c \in K$ is a constant then $f(P) = c$, and, for $f, g \in K[x_1, ..., x_n]$, define $(f+g)(P) = f(P) + g(P)$, $(fg)(P) = f(P)g(P)$. Then ψ_P is an onto ring homomorphism with $\mathrm{Ker}\psi_P = \langle x_1 - a_1, ..., x_n - a_n \rangle$ (check it!). By Proposition 1.2(ii), $\langle x_1 - a_1, ..., x_n - a_n \rangle$ is a maximal ideal of $K[x_1, ..., x_n]$. In Chapter 5 section 1 we will see that if K is algebraically closed then every maximal ideal of $K[x_1, ..., x_n]$ is of the form $\langle x_1 - b_1, ..., x_n - b_n \rangle$ for some $(b_1, ..., b_n) \in K^n$.

(iv) If p is a prime number in $\mathbb{Z}$, then $\langle 0 \rangle$, $\langle p \rangle$, $\langle x \rangle$ and $\langle p, x \rangle$ are prime ideals of $\mathbb{Z}[x]$, but only the last one is maximal. Indeed, the next proposition states a more general result.

1.3. Proposition Let R be a PID, K its field of fractions, and $R[x]$ the polynomial ring in x over R (hence a UFD). If $P \in \mathrm{Spec}R[x]$, then either

$P = \{0\}$, or $P = \langle f \rangle$ for some irreducible element $f \in R[x]$, or $P = \langle p, f \rangle$, where $p \in R$ is an irreducible element (hence a prime) and $f \in R[x]$ is a polynomial such that $\bar{f}$ is irreducible (hence a prime) in the polynomial ring $(R/\langle p \rangle)[x]$.

In the case where $P = \langle p, f \rangle$, $P \in \text{m-Spec}R[x]$.

Proof If $P = \langle 0 \rangle$ or $P = \langle f \rangle$ for some $0 \neq f \in R[x]$, then it is done by Example (ii) above.

Suppose P is not principal. Then there exist $f_1, f_2 \in P$ which do not have common divisor in $R[x]$. Note that R is a UFD. By Chapter 1 (section 2, exercise 7), f_1, f_2 do not have common divisor in $K[x]$, that is $\gcd(f_1, f_2) = 1$. Thus, $gf_1 + hf_2 = 1$ for some $g, h \in K[x]$. Multiplying by the common denominator, say u, of all coefficients in g and h, we have

$$(ug)f_1 + (uh)f_2 = u \in R \cap \langle f_1, f_2 \rangle \subset R \cap P.$$

This shows that $R \cap P \neq \{0\}$. But R is a PID. Hence $R \cap P = \langle p \rangle$ for some irreducible $p \in R$. By later exercise 2, the polynomial ring $\frac{R}{\langle p \rangle}[x]$ is a PID. Since $p \in P$, it follows from exercise 2 and the onto ring homomorphism

$$R[x] \longrightarrow \frac{R[x]}{\langle p \rangle} \xrightarrow{\cong} \frac{R}{\langle p \rangle}[x]$$

that if we pull the image of P in $\frac{R}{\langle p \rangle}[x]$ back into $R[x]$, then $P = \langle p, f \rangle$ as desired. □

Except for $\text{Spec}\mathbb{Z}[x]$, another typical case of Proposition 1.3 is $\text{Spec}K[x, y]$ where K is a field. In particular, it follows from the building of $\text{Spec}K[x, y]$ that all irreducible plane curves over K are established (see Chapter 5 section 2).

To indicate the existence of prime ideals in an arbitrary ring, we need Zorn's lemma that is equivalent to the axiom of choice in set theory.

1.4. Lemma (Zorn) Let Ω be a nonempty partially ordered set with the partial ordering $\succeq$. If any totally ordered subset $U \subset \Omega$ has an upper bound in Ω, then Ω contains a maximal element.

□

1.5. Proposition Let R be a ring.
(i) Any ideal $I \neq R$ is contained in a maximal ideal.

(ii) $R = U(R) \cup \left(\cup_{M \in \text{m-Spec}_R} M\right)$ where $U(R) \cap \left(\cup_{M \in \text{m-Spec}_R} M\right) = \emptyset$.

Proof (i) Set $\Omega = \{$proper ideals of R containing $I\}$ and let $\subseteq$ be the inclusion ordering on Ω. Then $I \in \Omega$ and $\subset$ is a partial ordering on Ω. If $V = \{J_i\}_{i \in \Lambda}$ is any totally ordered subset of Ω, then $J^* = \cup_{i \in \Lambda} J_i$ is an upper bound of V in Ω because $1 \notin J^*$. By Zorn's lemma, Ω contains a maximal element M. By the definition of Ω, M is also maximal among all ideals of R.
(ii) This follows from part (i) immediately. □

1.6. Proposition Let R be a ring.
(i) If S is a multiplicative set and I is an ideal of R with $I \cap S = \emptyset$, then there exists a prime ideal P such that

$$P \supseteq I \text{ and } P \cap S = \emptyset.$$

(ii) Every prime ideal contains a nonzero minimal prime ideal (with respect to $\subseteq$ on $\text{Spec}R$).

Proof (i) Consider the partially ordered set $\Omega = \{$ideals J with $J \supseteq I$ and $J \cap S = \emptyset\}$ with the inclusion ordering $\subseteq$. By Zorn's lemma, Ω contains a maximal element P. We claim that $P \in \text{Spec}R$. For if $a, b \in R$, $a, b \notin P$, then

$$(aR + P) \cap S \neq \emptyset, \quad (bR + P) \cap S \neq \emptyset,$$

because P is properly contained in both $aR + P$ and $bR + P$. Thus, there are $ax + p_1 \in (aR+P) \cap S$, $by + p_2 \in (bR+P) \cap S$, and $(ax+p_1)(by+p_2) = p' + abxy \in S$ since S is a multiplicative set, where $p' \in P$. This shows that $ab \notin P$ by the choice of P.
(ii) Exercise. □

Let R be a ring and $a \in R$. If $a^n = 0$ for some $n \geq 1$, then a is called a *nilpotent element*. Set

$$r(R) = \left\{a \in R \mid a \text{ is nilpotent}\right\}$$

and call $r(R)$ the *nilradical* of R. R is said to be *reduced* if $r(R) = \{0\}$.

1.7. Theorem (i) $r(R) = \cap_{P \in \text{Spec}R} P$. Consequently $r(R)$ is an ideal of R.

(ii) $r(R) = \cap Q$ where Q runs over all nonzero minimal prime ideals of R.

Proof (i) The inclusion $r(R) \subseteq \cap P$ is clear. The inclusion $\supseteq$ follows from the fact that if $a \in R$ is not nilpotent then there is some prime P not containing a. To see this, consider the multiplicative set $S = \{1, a, a^2, ..., a^n, ...\}$. Then $0 \notin S$, for a is not nilpotent. By Proposition 1.6(i) (taking $I = \{0\}$), there is a prime ideal P with $P \cap S = \emptyset$, as desired.
(ii) This follows from the proof of part (i) and Proposition 1.6(ii). $\square$

For a ring R, we also set the intersection

$$J(R) = \bigcap_{M \in \text{ m-Spec} R} M$$

and call $J(R)$ the *Jacobson radical* of R.

1.8. Theorem With notation as above,

$$J(R) = \left\{ r \in R \mid 1 - yr \in U(R) \text{ for all } y \in R \right\}.$$

Proof By Proposition 1.5(i), $1 - yr$ is a unit if and only if $1 - yr$ is not contained in any maximal ideal of R. Now the assertion is clear. $\square$

1.9. Theorem Let R be a Noetherian ring and $J(R)$ its Jacobson radical. If I is an ideal of R and $I \subseteq J(R)$, then $\cap_{n=1}^{\infty} I^n M = \{0\}$ holds for any finitely generated R-module M.

Proof This follows from Chapter 1 (section 7, Theorem 7.7 and exercise 6). $\square$

Exercises
1. Complete the proof of Proposition 1.6(ii). (Hint: Apply Zorn's lemma to the prime ideals contained in a prime ideal P by defining $P_2 \prec P_1$ if $P_1 \subset P_2$.)
2. Let R be a PID. Show that $\text{Spec} R = \text{m-Spec} R$.
3. Let R be a UFD.
   (a) Show that every minimal nonzero prime ideal of R is principal.
   (b) Show that, without counting associates, there is a one-to-one and onto correspondence between prime elements of R and minimal nonzero prime ideals of R.

4. Let R be a ring and I an ideal of R. Define
$$\sqrt{I} = \left\{ a \in R \ \middle|\ a^n \in I \text{ for some } n \geq 1 \right\}$$
and call $\sqrt{I}$ the radical of I in R. Show that
$$\sqrt{I} = \bigcap_{P \in \mathrm{Spec} R,\ P \supseteq I} P.$$

5. Let I and J be ideals of a ring R. Show that $\sqrt{I \cap J} = \sqrt{I} \cap \sqrt{J}$.

6. Let R be a ring. Two ideals I, J of R are said to be *comaximal* if $I + J = R$ (for instance, if one of them is maximal).
   (a) Show that if I and J are comaximal then $I \cdot J = I \cap J$, $I + J^2 = R$, and $I^m + J^n = R$ for all integers $m, n \geq 1$.
   (b) If $I_1, ..., I_N$ are ideals of R, and I_i and $J_i = \cap_{j \neq i} I_j$ are comaximal for all $i = 1, ..., N$, show that $I_1^n \cap \cdots \cap I_N^n = (I_1 \cdots I_N)^n = (I_1 \cap \cdots \cap I_N)^n$ for all integers $n \geq 1$.

7. Show that in $\mathbb{Z}[\sqrt{6}]$ the ideal $P = \langle 2, \sqrt{6} \rangle$ is a maximal ideal but not a principal ideal. (Hint: $\mathbb{Z}[\sqrt{6}]/P \cong \mathbb{Z}_2$)

8. For $f = 1 + x^2$, $g = y^2(x + x^3) + (y - 1)x^2 + y + 2$ in the polynomial ring $\mathbb{R}[x, y]$, is the ideal $\langle f, g \rangle$ a prime ideal or a maximal ideal?

2. Local Rings and DVRs

2.1. Definition Let R be a ring. If R has only one maximal ideal, then R is called a *local ring*.

By definition, all fields are local rings, for $\{0\}$ is the only maximal ideal in these rings.

2.2. Theorem The following statements are equivalent for a ring R.
(i) R is a local ring.
(ii) All nonunits of R form an ideal.

Proof This follows from Proposition 1.5. □

Example The following examples, which are from different aspects of mathematics, may perhaps help to qualify the name "local ring".

(i) (Foundation of modern scheme theory) Let V be the real line (or a topological space, or a differentiable manifold), and $P \in V$ a point. Consider the set E of real-valued continuous functions (differentiable functions in case V is a differentiable manifold) on some open interval (open neighborhood) around P. Then, two functions $f, g \in E$ may be "locally" identified by the equivalence relation: $f \sim g$ if and only if they agree on some open neighborhood of P. The quotient set $\mathcal{E}_P = (E \times E)/\sim$ forms a ring with the addition and multiplication induced by that on functions. Elements of $\mathcal{E}_P$ are called *function germs* at P. It is easy to see that the subset

$$\mathbf{m}_\mathbf{P} = \left\{ f \in \mathcal{E}_P \mid f(P) = 0 \right\}$$

forms an ideal of $\mathcal{E}_P$. If $g \notin \mathbf{m}$, then there is an open neighborhood U of P on which g is nonzero. So $h = \frac{1}{g}$ is defined on U and $gh = 1$ on U. By Theorem 2.2, $\mathcal{E}_P$ is a local ring.

(ii) Let $\mathbb{R}[x]$ be the polynomial ring in x over $\mathbb{R}$, and $\mathbb{R}(x)$ its field of fractions. For any $a \in \mathbb{R}$,

$$\mathbb{R}[x]_a = \left\{ \frac{f(x)}{g(x)} \in \mathbb{R}(x) \mid g(a) \neq 0 \right\}$$

is a local ring with the unique maximal ideal $\mathbf{m}_a = \{\frac{f(x)}{g(x)} \in \mathbb{R}(x) \mid g(a) \neq 0, \ f(a) = 0\}$. $\mathbb{R}[x]_a$ is called the local ring of the point a, due to the fact that each rational function in $\mathbb{R}[x]_a$ is "locally defined" on some Zariski open neighborhood of a. (See Chapter 5 section 3 about this topic in a more general setting.)

(iii) Let p be a prime in $\mathbb{Z}$. Then

$$\mathbb{Z}_{\langle p \rangle} = \left\{ \frac{a}{b} \in \mathbb{Q} \mid p \nmid b \right\}$$

is a local ring with the unique maximal ideal $\mathbf{m}_p = \{\frac{kp}{b} \in \mathbb{Z}_{\langle p \rangle} \mid k \in \mathbb{Z}\}$.

(iv) Let K be a field and R the ring of formal series in one variable, i.e.,

$$R = \left\{ \sum_{i=0}^{\infty} a_n x^i \mid a_i \in K \right\}.$$

If $f = a_0 + a_1 x + a_2 x^2 + \cdots$ with $a_0 \neq 0$, then $f = a_0(1 + xg)$ for some $g \in R$ and

$$a_0(1 + xg) \cdot a_0^{-1}(1 - xg + x^2 g^2 - \cdots) = 1.$$

Conversely, if $fh = 1$ for some $h \in R$, then $a_0 \neq 0$. Thus, R is a local ring with the unique maximal ideal $\mathbf{m} = xR = \langle x \rangle$.

The next easy but powerful lemma enables us to see the link between finitely generated modules over a local ring R and finite dimensional vector spaces over the field $R/\mathbf{m}$, where $\mathbf{m}$ is the unique maximal ideal of R.

2.3. Lemma (Nakayama) Let R be a ring and I an ideal contained in the Jacobson radical $J(R)$ of R (see section 1). If M is a finitely generated R-module and $IM = M$, then $M = \{0\}$.

Proof If $IM = M$ and $M \neq \{0\}$, then let $M = \sum_{i=1}^{s} R\xi_i$ for some $\xi_i \in M$, and we may assume that $\{\xi_1, ..., \xi_s\}$ is a minimal set of generators for M. Thus $\xi_i = \sum a_j \xi_j$ with $a_j \in I$, and hence $(1 - a_i)\xi_i = \sum_{j \neq i} a_j \xi_j$. But $1 - a_i$ is invertible. It follows that the set of generators $\{\xi_1, ..., \xi_s\}$ can be reduced, a contradiction. □

2.4. Corollary Let R be a local ring with the unique maximal ideal $\mathbf{m}$. Let M be an R-module and $N \subseteq M$ a submodule. Suppose that M/N is a finitely generated R-module and $M = N + \mathbf{m}M$. Then $N = M$.

In particular, let M be a finitely generated R-module. If $\xi_1, ..., \xi_s \in M$ are such that their images span the vector space $\overline{M} = M/\mathbf{m}M$ over the field $R/\mathbf{m}$, then $\xi_1, ..., \xi_s$ generate M.

Proof By Lemma 2.3, this follows from the fact that $J(R) = \mathbf{m}$. □

Remark If R is Noetherian, then Lemma 2.3 may follow from Proposition 1.9. In particular, if R is a Noetherian local ring with maximal ideal $\mathbf{m}$, then $\cap_{n=1}^{\infty} \mathbf{m}^n = \{0\}$.

A very important class of local rings from number theory and (analytic and algebraic) geometry is the class of discrete valuation rings, that is introduced as follows.

Let K be a field. Define in $\mathbb{Z} \cup \{\infty\}$ the rule $\infty + a = \infty$, $\infty + \infty = \infty$. A *discrete valuation* on K is an *onto* function

$$v : K \to \mathbb{Z} \cup \{\infty\}$$

$$x \mapsto v(x)$$

such that for all $x, y \in K$
(a) $v(0) = \infty$;
(b) $v(xy) = v(x) + v(y)$;
(c) $v(x+y) \geq \min\{v(x), v(y)\}$.

By definition, it is clear that $v(1_K) = 0 = v(-1_K)$. Hence $v(x) = v(-x)$ for all $x \in K$. Thus, if we set

$$R = \left\{ x \in K \mid v(x) \geq 0 \right\},$$

then it is easy to verify that R is a subring of K with $1_K \in R$.

2.5. Definition The ring R defined above is called the *discrete valuation ring* (abbreviated DVR) associated to the discrete valuation v.

2.6. Theorem Let R be a DVR associated to a discrete valuation v on a field K. Then R has the following properties.
(i) R is a local ring with the unique maximal ideal

$$\mathbf{m} = \left\{ x \in K \mid v(x) > 0 \right\}.$$

(ii) R is a PID, hence Noetherian. In particular, $\operatorname{Spec} R = \{\{0\}, \mathbf{m}\}$.

Proof (i) By the definition of v, $v(x^{-1}) = -v(x)$ for any $0 \neq x \in K$, and $x \in R$ is a unit if and only if $v(x) = 0$. This proves (i).
(ii) Let $t \in K$ be any element such that $v(t) = 1$. If $x \in \mathbf{m}$, then $v(x) = n > 1$, or equivalently, $v(x) = v(t^n)$. Hence $v(xt^{-n}) = 0$ and $xt^{-n} = u$ for some unit in R. Thus, $x = ut^n$, and every ideal I of R is of the form $\langle t^m \rangle$ for some $m \geq 1$. Therefore, R is a PID, and it is now clear that $\operatorname{Spec} R = \{\{0\}, \mathbf{m}\}$. □

2.7. Corollary A DVR R is a UFD. If R has maximal ideal $\mathbf{m} = \langle t \rangle$, then t (up to a unit multiple) is the unique prime in R. (In the literature, t is sometimes called a *uniformizing element* of R.)
□

Example (v) (Compare with previous Example (iii).) Let R be a UFD and K its field of fractions. Let p be a prime in R. Then for $0 \neq \frac{a}{b} \in K$,

$\frac{a}{b} = p^m \frac{a'}{b'}$ with a *unique* $m \in \mathbb{Z}$, and $p \nmid a'$, $p \nmid b'$. Thus

$$v: K \longrightarrow \mathbb{Z} \cup \{\infty\}$$

$$\frac{a}{b} \mapsto \begin{cases} m, & \text{if } 0 \neq \frac{a}{b} = p^m \frac{a'}{b'}, \ p \nmid a' \text{ and } p \nmid b', \\ \infty, & \text{if } \frac{a}{b} = 0 \end{cases}$$

defines a discrete valuation on K, called the *p-adic valuation* on K, where the associated valuation ring

$$V_p = \left\{ \frac{p^m a}{b} \in K \ \middle| \ m \geq 0, \ p \nmid b \right\}$$

has the unique maximal ideal

$$\mathbf{m} = \left\{ \frac{p^m a}{b} \in K \ \middle| \ m > 0, \ p \nmid b \right\} = \langle p \rangle.$$

(vi) The only discrete valuations on $\mathbb{Q}$ are p-adic valuations. To see this, let $v: \mathbb{Q} \to \mathbb{Z} \cup \{\infty\}$ be a nonzero discrete valuation on $\mathbb{Q}$ and R its valuation ring with the maximal ideal $\mathbf{m}$. Note that $\mathbb{Z} \subset R$ because $1 \in R$. Then

$$\mathbf{m} \cap \mathbb{Z} = \begin{cases} \{0\} \text{ or} \\ \langle p \rangle \text{ for some prime number } p \in \mathbb{Z}. \end{cases}$$

Since v is onto, it follows from the proof of Theorem 2.6(i) that $\mathbf{m} \cap \mathbb{Z} = \langle p \rangle$ for some prime number p. Thus, if $0 \neq \frac{a}{b} \in \mathbb{Q}$ then $\frac{a}{b} = p^k \left(\frac{a'}{b'} \right)$ with $k \in \mathbb{Z}$, $\gcd(p, a') = 1$, and $\gcd(p, b') = 1$. Hence $a', b' \notin \mathbf{m}$, or in other words, a' and b' are units of R. It follows that $v\left(\frac{a}{b}\right) = kv(p)$. Suppose $v(\lambda) = 1$. Then $v(\lambda) = mv(p)$ for some $m \in \mathbb{Z}$ implies $v(p) = 1$ because $p \in \mathbf{m}$. This shows that v is the p-adic valuation.

(vii) Let $K[x]$ be the polynomial ring in x over a field K and $K(x)$ its field of fractions. Consider all discrete valuations v on $K(x)$ such that $v(\lambda) = 0$ for $\lambda \in K^\times$. Then there is only one such v which is not a p-adic valuation. To see this, let $v: K(x) \to \mathbb{Z} \cup \{\infty\}$ be a nonzero discrete valuation on $K(x)$ and R its valuation ring with the maximal ideal $\mathbf{m}$. Below we consider two cases.

Case I $x \in R$. In this case $K[x] \subset R$ and

$$\mathbf{m} \cap K[x] = \begin{cases} \{0\} \text{ or} \\ \langle p(x) \rangle \text{ for some irreducible } p(x) \in K[x]. \end{cases}$$

Arguing as in Example (vi), we may conclude that v is a $p(x)$-adic valuation. Case II $x \notin R$. It follows from the proof of Theorem 2.6(i) that $y = \frac{1}{x} \in$ m. Hence, $K[y] \subset R$. Note that $K[y] \cong K[x]$ and $y \in K[y] \cap$ m. So $K[y] \cap$ m $= \langle y \rangle$ (why?). Thus, if $f = a_n x^n + \cdots + a_1 x + a_0$ with $a_n \neq 0$, then

$$f(x) = (a_n + a_{n-1} y + \cdots + a_1 y^{n-1} + a_0 y^n) y^{-n}$$

$$= q(y) y^{-n}, \text{ where } \gcd(q(y), y) = 1.$$

This shows that $v(q(y)) = 0$ and consequently $v(f(x)) = -nv(y)$. Now, for any $0 \neq \frac{f(x)}{g(x)} \in K(x)$, if $\deg f(x) = n$ and $\deg g(x) = m$, then $v\left(\frac{f(x)}{g(x)}\right) = (m-n)v(y)$. As in Example (v) we may derive from $y \in$ m that $v(y) = 1$. So eventually we have

$$v\left(\frac{f(x)}{g(x)}\right) = m - n,$$

that is, v is actually defined by the degree of polynomials.

2.8. Proposition Let R be a UFD and K its field of fractions. Then

$$R = \cap V_p$$

where V_p runs over all p-adic valuation rings in K.

Proof By the definition of a p-adic valuation on K, $R \subseteq \cap V_p$. If $0 \neq x \in \cap V_p$, say, $x = \frac{a}{b}$, $a, b \in R$, then $v(x) \geq 0$ for any prime p. Suppose

$$x = \frac{p_1^{\alpha_1} \cdots p_n^{\alpha_n}}{q_1^{\beta_1} \cdots q_m^{\beta_m}}$$

where p_i's and q_j's are primes and $\alpha_i, \beta_j > 0$. If some q_j does not appear in the numerator of x, then $v_{q_j}(x) < 0$; if $q_j = p_j$ but $\beta_j > \alpha_j$, then $v(q_j) < 0$. It follows that $m \leq n$, $p_j = q_j$ for $1 \leq j \leq m$ (up to a necessary re-arrangement of prime divisors), and $\beta_j \leq \alpha_j$ for $1 \leq j \leq m$. This shows that $x \in R$. Therefore $R \subseteq \cap V_p \subseteq R$. $\square$

To see under what condition a local domain is a DVR, we need the following result that is a special case of Chapter 1 Corollary 7.8.

2.9. Lemma Let R be a Noetherian domain and I an ideal of R. Then
$$\bigcap_{n=1}^{\infty} I^n = \{0\}.$$

$\square$

2.10. Theorem Let R be a local domain and $\mathbf{m}$ its maximal ideal. Suppose $\mathbf{m} = \langle t \rangle$ for some $t \neq 0$. If $\cap_{n=1}^{\infty} \mathbf{m}^n = \{0\}$, then the following statements hold:
(i) Every $0 \neq a \in R$ is of the form $a = t^n u$ for some $n \geq 0$ and some unit u.
(ii) Let K be the field of fractions of R. Define $v(a) = n$ for $a = t^n u$ as in part (i), and

$$K \longrightarrow \mathbb{Z} \cup \{\infty\}$$

$$\frac{a}{b} \mapsto \begin{cases} v(a) - v(b), & \text{if } \frac{a}{b} \neq 0, \\ \infty, & \text{if } \frac{a}{b} = 0. \end{cases}$$

Then v is a discrete valuation on K with R its associated valuation ring.
(iii) Every nonzero ideal I of R is of the form $I = \langle t^n \rangle$ for some $n \geq 0$.

Proof (i) By the assumption, if $0 \neq a \in R$ is not a unit then $a \in \mathbf{m}^\ell - \mathbf{m}^{\ell+1}$ for some $\ell \geq 1$, and it follows that $a = t^n u$ for some $n \geq 1$ and some unit.
(ii) By part (i), v is well-defined and surjective. Moreover, for $s \in K$, $v(s) \geq 0$ if and only if $s \in R$.

To verify that v is a discrete valuation on K, it is sufficient to note that for $a, b \in R$ with $b \neq 0$, if $v(a) \geq v(b)$ then $\frac{a+b}{b} = \frac{a}{b} + 1 \in R$ and $v(a+b) \geq v(b)$; if $v(a) > v(b)$ then $\frac{a+b}{b} = \frac{a}{b} + 1 \in R$ is a unit and $v(a+b) = v(b)$.
(iii) By part (i), let n be the smallest integer such that $t^u \in I$ where u is some unit. Then it is easy to see that $I = \langle t^n \rangle$. $\square$

A full structural characterization of a DVR is given in Chapter 3 Theorem 4.5.

Appendix. General valuation rings

Let R be the associated discrete valuation ring of a discrete valuation v on some field K. If $x \in K$, $x \notin R$, then $x \neq 0$ and $v(x^{-1}) > 0$, i.e., $x^{-1} \in R$.

In a general valuation theory (cf. [Jac], [Coh]), all (not necessarily discrete) valuation rings are defined in this way.

2.11. Definition Let K be a field and R a subring of K. If for every $x \in K^\times$, either x or $x^{-1} \in R$, then R is called a *valuation ring* of K.

If R is a valuation ring of some field K, then it is easy to see that $K = Q(R)$, that is, K is necessarily the field of fractions of R.

To see why the above definition is really more general than a discrete valuation ring, we need to introduce general valuation functions.

By an *ordered* abelian group $(G, +; \preceq)$ we mean an abelian group $(G, +)$ equipped with a *total ordering* $\preceq$ which is compatible with the binary operation $+$: $a \preceq b$ and $c \prec d$ implies $a + c \prec b + d$. For example, $(\mathbb{Z}, +; \leq)$, $(\mathbb{Q}, +; \leq)$ and $(\mathbb{R}, +; \leq)$ are ordered abelian groups.

Let K be a field and $(G, +; \preceq)$ an ordered abelian group. Define in $G \cup \{\infty\}$ the rule $\infty \succ a$, $\infty + a = \infty$, for all $a \in G$, and $\infty + \infty = \infty$. If there is an *onto* function

$$v: K \longrightarrow G \cup \{\infty\}$$
$$x \mapsto v(x)$$

such that
(a) $v(0) = \infty$,
(b) $v(xy) = v(x) + v(y)$ and
(c) $v(x + y) \succeq \min\{v(x), v(y)\}$,
then v is called a *valuation* on K.

2.12. Proposition Let v be a valuation on the field K. Then

$$R = \left\{ x \in K \mid v(x) \succeq 0 \right\}$$

is a valuation ring in K. Moreover, R is a local ring with maximal ideal $\mathbf{m} = \{x \in R \mid v(x) \succ 0\}$.

Proof Exercise. □

Let R be the valuation ring associated to a valuation v on the field K.

For $x, y \in K^\times$, suppose $v(x) = g$, $v(y) = h$. Then
$$\frac{y}{x} \in R \text{ if and only if } v\left(\frac{y}{x}\right) = v(y) - v(x) = h - g \succeq 0$$
if and only if $g \preceq h$.

This, indeed, induces a total order on the abelian group $K^\times$.

Motivated by the above idea, let R be a subring of the field K. If we write the binary operation of the abelian group $G = K^\times / U(R)$ additively, that is, $\overline{x} + \overline{y} = \overline{x} \cdot \overline{y}$, for $x, y \in K^\times$, then a partial ordering $\preceq$ on G may be defined as
$$\overline{x} \preceq \overline{y} \text{ if and only if } \frac{y}{x} \in R, \; x, y \in K^\times.$$

2.13. Proposition Let $R \subset K$ and $(G, +; \preceq)$ be as above. Then the following hold:

(i) The correspondence
$$v : K \longrightarrow G \cup \{\infty\}$$
$$x \mapsto v(x) = \begin{cases} \infty, & \text{if } x = 0; \\ \overline{x}, & \text{if } x \neq 0, \end{cases}$$
defines a valuation function with R as its associated valuation ring if and only if $\preceq$ is a total ordering on G.

(ii) If $(H, +; \trianglelefteq)$ is some ordered abelian group and $v \colon K \to H \cup \{\infty\}$ is a valuation with the associated valuation ring R, then
$$\varphi : G \xrightarrow{\cong} H$$
as ordered abelian groups, i.e., $g \preceq h$ implies $\varphi(g) \trianglelefteq \varphi(h)$ for $g, h \in G$.

Proof Exercise. □

2.14. Theorem Let R be a valuation ring. Then R is a DVR if and only if R is Noetherian.

Proof One direction is known by Theorem 2.6. Suppose R is Noetherian. Then the maximal ideal $\mathbf{m}$ of R is finitely generated, say $\mathbf{m} = \langle a_1, a_2, ..., a_n \rangle$ with all $a_i \neq 0$. Thus, we may assume $\frac{a_1}{a_2} \in R$, and so $\mathbf{m} = \langle a_2, a_3, ..., a_n \rangle$. Repeating this procedure will finally yield $\mathbf{m} = \langle a_i \rangle$, a principal ideal. Hence R is a DVR by Theorem 2.10.

Exercises

1. Let $K[x]$ be the polynomial ring in x over a field K. Show that $R = K[x]/\langle x^n \rangle$ is a local ring for any $n \in \mathbb{Z}^+$. What is the maximal ideal of R?
2. Let R be a ring and M a maximal ideal of R. Show that R is a local ring with the unique maximal ideal M if and only if $1 + a$ is a unit in R for every $a \in M$.
3. Show that Nakayama's lemma holds for any ideal $I \subseteq r(R)$.
4. Consider the local ring V_p given in previous Example (v). Show that $K = V_p + \frac{1}{p}V_p + \frac{1}{p^2}V_p + \cdots$, and that $K = pK$. But $K \neq \{0\}$. This illustrates that Nakayama's lemma doesn't work if the module considered is not finitely generated. (See also Chapter 1 (section 7, exercise 3).)
5. Let R be a local ring and $\mathbf{m}$ its maximal ideal. If $\mathbf{m} = \langle t \rangle$ is principal, show that either R is a DVR or $t^m = 0$ for some $m > 1$. (Hint: Similar to the proof of Theorem 2.10.)
6. Complete the proof of Propositions 2.12 and 2.13.

3. The Ring of Fractions and Localization

Recall from (section 2, Examples (ii)–(iii)) that in the local ring

$$\mathbb{Z}_{\langle p \rangle} = \left\{ \frac{a}{b} \in \mathbb{Q} \ \Big| \ p \nmid b \right\},$$

all $b \in S_1 = \mathbb{Z} - \langle p \rangle$, and in the local ring

$$\mathbb{R}[x]_a = \left\{ \frac{f(x)}{g(x)} \in \mathbb{R}(x) \ \Big| \ g(a) \neq 0 \right\},$$

all $g(x) \in S_2 = \mathbb{R}[x] - \langle x - a \rangle$.

Observe that

- every $s \in S_1$ is invertible in $\mathbb{Z}_{\langle p \rangle}$, and every $s' \in S_2$ is invertible in $\mathbb{R}[x]_a$.

We may say that $\mathbb{Z}_{\langle p \rangle}$ is the ring of fractions of $\mathbb{Z}$ with denominators in S_1, and that $\mathbb{R}[x]_a$ is the ring of fractions of $\mathbb{R}[x]$ with denominators in S_2.

Also observe that

- the set of denominators S_1, respectively S_2, is closed under multiplication in $\mathbb{Z}$, respectively in $\mathbb{R}[x]$.

Considering any domain R and its field of fractions K, if S is a multiplicative set of R and $0 \notin S$ (for instance, $S = \{1, r, r^2, ...\}$, $0 \neq r \in R$), then a direct verification shows that

$$R_S = \left\{ \frac{a}{s} \,\middle|\, a \in R,\ s \in S \right\} \subset K$$

is a subring of K containing R. We may call R_S the ring of fractions with denominators in S.

Our aim in this section is to demonstrate that the above described idea of establishing the ring of fractions may be carried out in a more general setting including the case where the ring R may have divisors of zero.

Let R be a ring and S a multiplicative set in R. Define on $R \times S$ the relation

$(*)$ $\quad (a_1, s_1) \sim (a_2, s_2)$ if and only if there is $s \in S$ such that $(s_1 a_2 - s_2 a_1)s = 0$.

Then one easily sees that $\sim$ is *reflexive, symmetric*. If $(a_1, s_1) \sim (a_2, s_2)$ and $(a_2, s_2) \sim (a_3, s_3)$, then

$$\begin{cases} (s_1 a_2 - s_2 a_1)s = 0 \\ (s_2 a_3 - s_3 a_2)s' = 0 \end{cases} \text{implies} \quad \begin{cases} (s_1 a_2 - s_2 a_1)s s_3 s' = 0 \\ (s_2 a_3 - s_3 a_2)s' s_1 s = 0 \end{cases}$$

and thus, $0 = s_2 a_3 s' s_1 s - s_2 a_1 s s_3 s' = (s_1 a_3 - s_3 a_1)s s_2 s'$, i.e., $(a_1, s_1) \sim (a_3, s_3)$. This shows that $\sim$ is also *transitive*, and hence $\sim$ is an equivalence relation on $R \times S$.

Write R_S for the quotient set $R \times S / \sim$, and write $\frac{a}{s}$ for the equivalence class of $(a, s) \in R \times S$. Define the addition and multiplication on R_S as follows:

$$\frac{a_1}{s_1} + \frac{a_2}{s_2} = \frac{s_2 a_1 + s_1 a_2}{s_1 s_2}, \quad \frac{a_1}{s_1} \cdot \frac{a_2}{s_2} = \frac{a_1 a_2}{s_1 s_2}.$$

Then a direct verification shows that R_S is an associative commutative ring with zero $0 = \frac{0}{1}$ and identity $1 = \frac{1}{1}$.

3.1 Definition The ring R_S constructed above is called the *ring of fractions* with denominators in S.

Remark Recall that if R is a domain and $S = R - \{0\}$, then the field K of fractions of R is given by $R \times S/\sim$ where $\sim$ is defined on $R \times S$ by

$(**)$ $\qquad (a_1, s_1) \sim (a_2, s_2)$ if and only if $s_1 a_2 - s_2 a_1 = 0$.

The reason why we did not define $\sim$ in the foregoing $(*)$ exactly as in $(**)$ above will be clear if one checks the transitivity of $\sim$ defined in $(**)$. (Also see Example (i) below.)

3.2. Proposition Let R_S be the ring of fractions with denominators in S.
(i) If $0 \in S$, then R_S is the zero ring.
(ii) The correspondence

$$\lambda_R : R \longrightarrow R_S$$

$$x \mapsto \frac{x}{1}$$

defines a ring homomorphism (called the *canonical homomorphism*) such that
(a) $\lambda_R(s) = \frac{s}{1}$ is invertible in R_S for every $s \in S$; and
(b) $\mathrm{Ker} \lambda_R = \{a \in R \mid as = 0 \text{ for some } s \in S\}$.
Hence λ_R is injective if and only if S does not contain nontrivial divisor of zero (note that this is compatible with the case where R is a domain).
(iii) λ_R has the universal property: If $f \colon R \to R'$ is any ring homomorphism such that $f(1_R) = 1_{R'}$ and $f(s)$ is invertible in R' for every $s \in S$, then there exists a *unique* ring homomorphism $\overline{f} \colon R_S \to R'$ with $\overline{f}\left(\frac{r}{s}\right) = \frac{f(r)}{f(s)}$ that yields the commutative diagram

$$\begin{array}{ccc} R & \xrightarrow{\lambda_R} & R_S \\ {\scriptstyle f} \downarrow & \swarrow \overline{f} & \\ R' & & \end{array} \qquad \overline{f} \circ \lambda_R = f$$

Proof Exercise. □

Example (i) Let $K[x, y]$ be the polynomial ring in x and y over a field K. Consider the ring $R = K[x, y]/\langle xy \rangle$ and $S = \{1, \overline{x}, \overline{x}^2, ...\}$. Then
(a) the previous $(**)$ itself cannot define an equivalence relation on $R \times S$; and
(b) since $\overline{xy} = 0$ and $\overline{x}$ is invertible in R_S, it follows that $R_S \cong K[t, t^{-1}] \subset K(t)$, where $K[t]$ is the polynomial ring in t over K.

Note that R has nontrivial divisors of zero, but R_S is a domain.

(ii) Let R be a ring and $f \in R$. Consider $S = \{1, f, f^2, ...\}$. Then

$$R_S \cong \frac{R[x]}{\langle xf - 1 \rangle},$$

where $R[x]$ is the polynomial ring in x over R. (This trick plays a very important role in algebraic geometry, for example, in proving the Nullstellensatz and in obtaining the open affine covering of an algebraic set, etc.)

Proof Note that $\overline{f}$ is invertible in $R[x]/\langle xf - 1 \rangle$. By the universal property of R_S, the diagram

$$\begin{array}{ccc} R & \xrightarrow{\lambda_R} & R_S \\ {\scriptstyle \alpha} \downarrow & \nearrow {\scriptstyle \bar{\alpha}} & \\ \dfrac{R[x]}{\langle xf - 1 \rangle} & & \end{array}$$

is commutative, where α is the natural ring homomorphism $R \to R[x]/\langle xf - 1 \rangle$. On the other hand, under the ring homomorphism $\beta \colon R[x] \to R_S$ with $\beta(\sum r_i x^i) = \sum r_i \frac{1}{f^i}$, $\beta(xf - 1) = 0$. By the first isomorphism theorem (or see section 0), there is a ring homomorphism $\overline{\beta} \colon R[x]/\langle xf - 1 \rangle \to R_S$ that yields the commutative diagram

$$\begin{array}{ccc} R[x] & \xrightarrow{\pi} & \dfrac{R[x]}{\langle xf - 1 \rangle} \\ {\scriptstyle \beta} \downarrow & \swarrow {\scriptstyle \bar{\beta}} & \\ R_S & & \end{array} \qquad \overline{\beta} \circ \pi = \beta$$

Now one checks that $\overline{\alpha}$ and $\overline{\beta}$ are inverses to each other.

The next proposition will be generalized in Chapter 3 Theorem 4.1.

3.3. Proposition Let R be a UFD, S a multiplicative set of R and $0 \notin S$. Then R_S is a UFD.

Proof View R as a subring via the canonical mapping $\lambda_R \colon R \to R_S$. We show that factorization in R_S is completely determined by that in R. First note that if p is a prime in R then $p|s$, for some $s \in S$, if and only if p is a unit in R_S. Secondly, note that if p is a prime of R but not a unit in R_S

then p is a prime in R_S. (Indeed, using the factorization in R it is easy to see that every prime in R_S is of the form pu for some prime p of R and some unit in R_S). Finally, note that if $\frac{r}{s} \in R_S$ with $r \neq 0$ then

$$\frac{r}{s} = p_1^{\alpha_1} p_2^{\alpha_2} \cdots p_m^{\alpha_m} \cdot u,$$

where the p_i are primes of R but not units in R_S while u is a unit in R_S. It follows that R_S is a UFD by Chapter 1 Theorem 2.9. □

Let R be a ring, R_S the ring of fractions of R with denominators in a multiplicative set S, and $\lambda_R \colon R \to R_S$ the canonical ring homomorphism. We now consider the relation between ideals of R and ideals of R_S.

If I is an ideal of R, then

$$I^e = \lambda_R(I) R_S = \left\{ \frac{a}{s} \;\middle|\; a \in I,\ s \in S \right\}$$

is an ideal of R_S and is called the *extension* of I in R_S. If J is an ideal of R_S, then the preimage

$$J^c = \lambda_R^{-1}(J) = \left\{ r \in R \;\middle|\; \lambda(r) \in J \right\}$$

is an ideal of R and is called the *contraction* of J in R.

3.4. Proposition With notation as above, the following hold:
(i) For any ideal $J \subseteq R_S$, $J^{ce} = J$.
(ii) For any ideal $I \subseteq R$,

$$I^{ec} = \left\{ r \in R \;\middle|\; rs \in I \text{ for some } s \in S \right\}.$$

(iii) If P is a prime ideal and $P \cap S = \emptyset$, then P^e is a prime ideal of R_S.

Proof (i) If $\frac{r}{s} \in J$ then $r \in J^c$, and hence $\frac{r}{s} \in J^{ce}$. This shows that $J \subseteq J^{ce}$. The inclusion $J^{ce} \subseteq J$ is clear.
(ii) If $r \in I^{ec}$ then $\frac{r}{1} = \frac{r'}{s} \in R_S$ for some $r' \in I$ and $s \in S$. Thus, $s'sr = s'r' \in I$ for some $s' \in S$. This proves the inclusion $\subseteq$. The inclusion $\supseteq$ is again clear.
(iii) This follows from (ii) immediately. □

3.5. Corollary (i) Let I be an ideal of R. Then $I^{ec} = I$ if and only if

(•) $\qquad\qquad rs \in I$ implies $r \in I$ for $r \in R$ and $s \in S$.

(ii) There is a one-to-one and onto correspondence defined by extension and contraction of ideals:

$$\{\text{ideals of } R \text{ satisfying } (\bullet) \text{ in part (i)}\} \leftrightarrow \{\text{ideals of } R_S\}.$$

It follows that if R is Noetherian then so is R_S. In particular, if R is a principal ideal ring then so is R_S.

(iii) $I^{ec} = R$ if and only if $I^e = R_S$ if and only if $I \cap S \neq \emptyset$.

(iv) If $P \in \text{Spec} R$ such that $P \cap S = \emptyset$, then $(*)$ holds for P, and $P^{ec} = P$. It follows that there is a one-to-one and onto correspondence defined by extension and contraction of ideals:

$$\left\{ P \in \text{Spec} R \mid P \cap S = \emptyset \right\} \leftrightarrow \text{Spec} R_S.$$

Proof Exercise. □

Let R be a ring and $P \in \text{Spec} R$. Then $S_P = R - P$ is a multiplicative set of R. Write R_P for the ring of fractions of R with denominators in S_P.

3.6. Proposition With notation as above, the following hold:
(i) R_P is a local ring with the unique maximal ideal $\mathbf{m} = PR_P = P^e$.
(ii) There is a one-to-one and onto correspondence defined by extension and contraction of ideals:

$$\left\{ Q \in \text{Spec} R \mid Q \subseteq P \right\} \leftrightarrow \text{Spec} R_P.$$

Proof Note that $\frac{r}{s} \in R_P$ is a nonunit if and only if $r \in P$. So (i) holds.
(ii) follows from Corollary 3.5(iv). □

3.7. Definition With notation as above, R_P is called the *localization* of R at P.

The above definition comes from the "local study" of points in algebraic geometry (see (section 2, Example (ii)) and Chapter 5 section 3). For instance, if $R = K[x_1, ..., x_n]$ and K is algebraically closed, then the localization R_M of R at a maximal ideal M is the ring of all rational functions in $K(x_1, ..., x_n)$ that are defined on some (open) neighborhood of the point determined by M (Chapter 5 Proposition 3.6). Furthermore, the next theorem says that the set of "global rational functions" that are defined at every point is nothing but the ring R (Chapter 5 Theorem 3.4).

3.8. Theorem Let R be a domain. Then

$$R = \cap R_M = \cap R_P,$$

where M runs over all maximal ideals of R, and P runs over all prime ideals of R.

Proof The inclusion $R \subseteq \cap R_P$ is clear. Let $x \in \cap R_P$, and set $I_x = \{a \in R \mid ax \in R\}$. Then it is easy to see that I_x is an ideal of R and $x \in R$ if and only if $1 \in I_x$. If $1 \notin I_x$, then there is a prime ideal P with $P \supseteq I_x$ by Proposition 1.5. But then $x \notin R_P$, for if $x = \frac{a}{b} \in R_P$, $b \notin P$, then $bx \in R$ and $b \in I_x \subseteq P$. This shows that $x \in R$ and hence $R = \cap R_P$.

Exercises

1. For $a, a_1, a_2 \in R$, $s, s' \in S$, verify in R_S

$$\frac{s'a}{s's} = \frac{a}{s}, \quad \frac{a_1 + a_2}{s} = \frac{a_1}{s} + \frac{a_2}{s}.$$

2. Let R be a domain and S a multiplicative set of R, where $0 \notin S$. Show that R and R_S have the same field of fractions.
3. Complete the proof of Proposition 3.2.
4. Let S be a multiplicative set in $\mathbb{Z}$, $0 \notin S$. Find all primes in $\mathbb{Z}_S$.
5. Give a detailed proof of Corollary 3.5.
6. Let $R = K[x,y]/\langle xy \rangle$ be as in previous Example (i).
   (a) For every $c \in K$, show that $P = \langle \overline{x} - c, \overline{y} \rangle$ is a maximal ideal of R.
   (b) For every P in part (a) with $c \neq 0$, show that R_P is a DVR. What happens if $c = 0$? (Hint: $\overline{y} = \frac{1}{c}(\overline{x} - c)\overline{y}$.)
7. Let $R = K[x,y]/\langle y^2 - x^3 \rangle$, where $K[x,y]$ is the polynomial in x,y over a field K. Show that R_P is not a DVR, where $P = \langle \overline{x}, \overline{y} \rangle$. (Hint: See Chapter 3 (section 3, Example (iii)).)
8. Let R be a PID. Show that if $P \in \text{Spec} R$ then R_P is a DVR.

4. The Module of Fractions

Let R be a ring and R_S the ring of fractions of R with denominators coming from a multiplicative set S, where $0 \notin S$. In this section we consider the relation between modules over R and modules over R_S.

Let $\overline{M}$ be an R_S-module. Then $\overline{M}$ also has an R-module structure via the canonical mapping $\lambda_R \colon R \to R_S$:

$$r \cdot x = \lambda_R(r)x = \frac{r}{1}x, \; r \in R, \; x \in \overline{M}.$$

In view of Chapter 1 section 7, there is a ring homomorphism $\rho \colon R \to \operatorname{End}_{\mathbb{Z}}\overline{M}$, where $\rho(r) = \rho_r$ with $\rho_r(y) = r \cdot y$ for $y \in \overline{M}$. Moreover, we observe that

- for every $s \in S$, ρ_s is invertible in $\operatorname{End}_{\mathbb{Z}}\overline{M}$, that is, every ρ_s is an isomorphism.

Conversely, let M be an R-module. Then there is a ring homomorphism $\varphi \colon R \to \operatorname{End}_{\mathbb{Z}}M$, where $\varphi(r) = \rho_r$ with $\rho_r(m) = rm$ for $r \in R$ and $m \in M$. Suppose that the property mentioned in the above ($\bullet$) holds for M, i.e., ρ_s is invertible for all $s \in S$. Then, by the universal property of R_S, there is a unique ring homomorphism $\overline{\varphi} \colon R_S \to \operatorname{End}_{\mathbb{Z}}M$ that yields the commutative diagram

$$\begin{array}{ccc} R & \xrightarrow{\lambda_R} & R_S \\ {\scriptstyle \varphi} \downarrow & \swarrow {\scriptstyle \overline{\varphi}} & \\ \operatorname{End}_{\mathbb{Z}}M & & \end{array} \qquad \overline{\varphi} \circ \lambda_R = \varphi$$

Consequently, M is equipped with an R_S-module structure.

The above remark leads to the following construction.

Let M be an R-module. Define the equivalence relation $\sim$ on $M \times S$ (check it!):

$(m_1, s_1) \sim (m_2, s_2)$ if and only if $s(s_1 m_2 - s_2 m_1) = 0$ for some $s \in S$,

and set $M_S = M \times S/\sim$. Write $\frac{m}{s}$ for the equivalence class of $(m, s) \in M \times S$, and then define the R-module structure by

$$\frac{m_1}{s_1} + \frac{m_2}{s_2} = \frac{s_2 m_1 + s_1 m_2}{s_1 s_2}, \quad r \cdot \frac{m}{s} = \frac{rm}{s}.$$

A direct verification shows that M_S is an R-module and every $s \in S$ defines an isomorphism ρ_s in $\operatorname{End}_{\mathbb{Z}}M_S$. It follows from the foregoing remark that M_S is an R_S-module on which the R_S-action is given by

$$\frac{r}{s_1} \cdot \frac{m}{s_2} = \frac{rm}{s_1 s_2}.$$

4.1. Definition The R_S-module M_S constructed above is called the *module of fractions* with denominators in S.
If $S = R - P$ for some $P \in \operatorname{Spec} R$, then M_S is called the *localization of M at P* and is denoted by M_P.

4.2. Proposition Let M be an R-module and M_S the R_S-module of fractions with denominators in the multiplicative set S.
(i) If $0 \in S$, then M_S is the zero module.
(ii) The correspondence

$$\mu_M : M \longrightarrow M_S$$

$$m \mapsto \frac{m}{1}$$

defines an R-module homomorphism (called the *canonical homomorphism*) with $\operatorname{Ker}\mu_M = \{m \in M \mid sm = 0 \text{ for some } s \in S\}$.
(iii) μ_M has the universal property: If $\overline{M}$ is an R_S-module and $f\colon M \to \overline{M}$ is an R-module homomorphism, then there is a unique R_S-module homomorphism $\overline{f}\colon M_S \to \overline{M}$ with $\overline{f}\left(\frac{m}{s}\right) = \frac{1}{s} \cdot f(m)$ that yields the commutative diagram

$$\begin{array}{ccc} M & \xrightarrow{\mu_M} & M_S \\ {\scriptstyle f}\downarrow & \swarrow {\scriptstyle \overline{f}} & \\ \overline{M} & & \end{array} \qquad \overline{f} \circ \mu_M = f$$

Proof Exercise. □

Let $\psi\colon M \to N$ be an R-module homomorphism with $K = \operatorname{Ker}\psi$ and $W = \operatorname{Im}\psi$. Then K is a submodule of M and W is a submodule of N. It is easy to check that

$$\psi_S : M_S \longrightarrow N_S$$

$$\frac{m}{s} \mapsto \frac{\psi(m)}{s}$$

is an R_S-module homomorphism.

4.3. Theorem With notation as above, the following hold:
(i) $\operatorname{Ker}\psi_S = K_S$, $\operatorname{Im}\psi_S = W_S$.

(ii) If ψ is injective then so is ψ_S; if ψ is onto then so is ψ_S.
(iii) $(M/K)_S \cong M_S/K_S$.

Proof (i)–(ii) It is clear that $K_S \subseteq \mathrm{Ker}\psi_S$. If $\psi_S\left(\frac{m}{s}\right) = \frac{\psi(m)}{s} = 0$, then $s'\psi(m) = 0$ for some $s' \in S$. Since ψ is an R-module homomorphism, $0 = s'\psi(m) = \psi(s'm)$, that is, $s'm \in \mathrm{Ker}\psi = K$. It follows that $\frac{m}{s} = \frac{s'm}{s's} \in K_S$. This shows that $K_S = \mathrm{Ker}\psi_S$. Similarly, one checks that $\mathrm{Im}\psi_S = W_S$.
(iii) This is just a consequence of applying parts (i)–(ii) to the natural onto R-module homomorphism $M \to M/K$. $\square$

Let R be a ring and I an ideal of R. In view of $R \xrightarrow{\lambda_R} R_S$ and Theorem 4.3, one may verify that $I_S = IR_S$, and moreover, that the rule $\frac{r}{s} \cdot \frac{r'}{s'} = \frac{rr'}{ss'}$ defines a ring multiplication on the R_S-module $(R/I)_S$.

4.4. Corollary (i) Let I be an ideal of R. Then there is a ring isomorphism $R_S/I_S \cong (R/I)_S$.
(ii) If $P \in \mathrm{Spec} R$, then $K = R_P/P_P$ is isomorphic to the field of fractions of the domain R/P.

Proof This follows from Theorem 4.3 and the above remark.

Exercises

1. Verify that the relation $\sim$ on $M \times S$ is an equivalence relation.
2. Verify that the R_S-module structure on M_S is well-defined.
3. For $m, m_1, m_2 \in M$, $s, s' \in S$, verify in M_S
$$\frac{s'm}{s's} = \frac{m}{s}, \quad \frac{m_1 + m_2}{s} = \frac{m_1}{s} + \frac{m_2}{s}.$$
4. Complete the proof of Proposition 4.2.
5. Consider the prime ideal $P = \langle p \rangle$ in $\mathbb{Z}$, where p is a prime number. Let $m \in \mathbb{Z}^+$, $m = p_1^{\alpha_1} p_2^{\alpha_2} \cdots p_m^{\alpha_m}$ the factorization of m into primes, and $M = \mathbb{Z}/\langle m \rangle$.
   (a) If $p \nmid m$, show that $M_P = \{0\}$.
   (b) If $p = p_i$ for some i, show that $M_P \cong \mathbb{Z}/\langle p_i^{\alpha_i} \rangle$. (Hint: Write $m = p_i^{\alpha_i} m_1$. Then $m_1 \in \mathbb{Z} - P$. Moreover, if $k \in \mathbb{Z}$, then $k = p_i^{\alpha_i} q + r$, $0 \le r < p_i^{\alpha_i}$.)
6. Let S be a multiplicative set of the ring R, and let M be a finitely generated R-module. Show that M_S is a finitely generated R_S-module. (Hint: Apply exercise 3 above to a direct verification, or use Proposition 4.3 and Chapter 1 Proposition 7.3.)

7. Let S be a multiplicative set of the ring R, and let I be an ideal of R. Viewing I as an R-module, show that $I_S = I^e$ via $I_S \hookrightarrow R_S$, where I^e is the extension of I in R_S in the sense of section 3. Furthermore, if J is another ideal of R, verify that the following hold:
   (a) $(I + J)_S = I_S + J_S$.
   (b) $(IJ)_S = I_S J_S$.
   (c) $(I \cap J)_S = I_S \cap J_S$.
8. State and verify a module-version of properties (a) and (c) above for submodules.

Chapter 3
Integral Extensions and Normalization

1. Integral Extensions

In commutative algebra a natural generalization of field extension is the ring extension. Note that field extensions may be studied in terms of vector space. Here the structure of an R-algebra is convenient for the study of ring extensions.

Let R and B be rings. B is said to be an *R-algebra* if there is a *nonzero* ring homomorphism $\varphi\colon R \to B$. Recalling the convention $\varphi(1_R) = 1_B$, it follows that B forms an R-module with the module structure:

$$(*) \qquad ab = \varphi(a)b, \ a \in R, \ b \in B.$$

If R is a field, then $R \cong \varphi(R) \subset B$ and B becomes an R-vector space.

An *R-subalgebra* of B is a subring $B' \subseteq B$ with $1_{B'} = 1_B$, which is also an R-module with respect to the action defined by the above $(*)$.

If B is an R-algebra, then it is clear that any ideal of B is an R-module.

Let B_1 and B_2 be R-algebras. An *R-algebra homomorphism* $\psi\colon B_1 \to B_2$ is simultaneously a ring homomorphism and an R-module homomorphism.

Let B be an R-algebra defined by the ring homomorphism $\varphi\colon R \to B$. If φ is a *ring monomorphism*, then R is naturally viewed as an R-subalgebra of B. In this case we say that B is an *extension ring* or *extension algebra* of R. Bearing this in mind, if no confusion arises, we simply refer $R \subseteq B$ to a *ring extension*.

The polynomial ring $R[x_1, ..., x_n]$ in variables $x_1, ..., x_n$ over a ring R is called a *polynomial R-algebra*.

1.1. Definition Let $R \subseteq B$ be a ring extension.
(i) If $B = R[S]$ for some nonempty subset $S \subset B$, where $R[S]$ is the subring of B generated by S over R (see Chapter 1 section 0 for the definition), we say that B is an *R-algebra generated by S* and call S a *set of generators* of B or a *generating set* of B. If $S = \{b_1, ..., b_n\}$ is a finite set, then B is called a *finitely generated R-algebra* and is denoted by $B = R[b_1, ..., b_n]$. In the latter case, we also say that B is a *finitely generated extension ring (algebra)* of R.
(ii) If B is a finitely generated R-module, then we say that B is a *module-finite extension ring (algebra)* of R or B is *module-finite* over R.

From definition it is clear that *any module-finite extension ring (algebra) is a finitely generated extension ring (algebra)*.

Example (i) Let R be a ring and $B = R[x_1, ..., x_n]$ the polynomial ring in variables $x_1, ..., x_n$ over R. Then B is a finitely generated extension ring (algebra) of R but B is not module-finite over R.

(ii) Let $d \in \mathbb{Z}$ be square-free (i.e., $n^2 \nmid d$ for any $n \in \mathbb{Z}$). Then $\mathbb{Z} \subset \mathbb{Z}[\sqrt{d}] = \mathbb{Z} + \mathbb{Z}\sqrt{d}$ is a module-finite extension.

(iii) Using the fact that $\mathbb{Z}$ is a UFD and there are infinitely many primes in $\mathbb{Z}$, we conclude that $\mathbb{Q}$ is not a finitely generated extension ring of $\mathbb{Z}$. (See also Chapter 1 (section 7, exercise 3).)

1.2. Definition (Compare with an algebraic element in a field extension.) Let B be an R-algebra defined by the ring homomorphism $\varphi\colon R \to B$. Write $R' = \varphi(R)$. An element $b \in B$ is said to be *integral* over R if there is a monic polynomial $f(x) \in R'[x]$, say $f(x) = x^n + \lambda_{n-1}x^{n-1} + \cdots + \lambda_1 x + \lambda_0$, $\lambda_i \in R'$, such that

$$f(b) = b^n + \lambda_{n-1}b^{n-1} + \cdots + \lambda_1 b + \lambda_0 = 0.$$

If every $b \in B$ is integral over R, we say that B is *integral* over R. If $R \subseteq B$ and B is integral over R, then we call B an *integral extension ring (algebra)* of R and refer $R \subseteq B$ to an *integral extension*.

1.3. Theorem Let B be an R-algebra defined by the ring homomorphism

$\varphi\colon R \to B$. Write $R' = \varphi(R)$. For $b \in B$, the following are equivalent:
(i) b is integral over R.
(ii) The R-subalgebra $R'[b] \subseteq B$ is a finitely generated R'-module (or equivalently, a finitely generated R-module).
(iii) There exists an R-subalgebra $B' \subseteq B$ with the property that $R'[b] \subseteq B'$ and B' is a finitely generated R'-module (or equivalently, a finitely generated R-module).

Proof (i) $\Rightarrow$ (ii) b is integral over R implies

$$b^n + \lambda_{n-1}b^{n-1} + \cdots + \lambda_1 b + \lambda_0 = 0, \ \lambda_i \in R'.$$

Inductively we derive that

$$b^{n+i} \in \sum_{j=0}^{n-1} R'b^j, \ i \in \mathbb{N}.$$

It follows that $R'[b] = \sum_{j=0}^{n-1} R'b^j$.
(ii) $\Rightarrow$ (iii) Take $B' = R'[b]$.
(iii) $\Rightarrow$ (i) Let $R'[b] \subseteq B' \subseteq B$, where B' is the R-subalgebra as described. Suppose $B' = \sum_{i=1}^n R'\xi_i$, $\xi_i \in B'$. Then $b\xi_i \in B'$, say

$$b\xi_i = \sum_{j=1}^n \lambda_{ij}\xi_j, \ \lambda_{ij} \in R', \ i = 1,...,n.$$

In a matrix form we have

$$\begin{pmatrix} \lambda_{11} - b & \lambda_{12} & \cdots & \lambda_{1n} \\ \lambda_{21} & \lambda_{22} - b & \cdots & \lambda_{2n} \\ \vdots & \vdots & \cdots & \vdots \\ \lambda_{n1} & \lambda_{n2} & \cdots & \lambda_{nn} - b \end{pmatrix} \begin{pmatrix} \xi_1 \\ \xi_2 \\ \vdots \\ \xi_n \end{pmatrix} = \begin{pmatrix} 0 \\ 0 \\ \vdots \\ 0 \end{pmatrix}.$$

Write M for the above $n \times n$ matrix. If we multiply the above equality by the adjoint matrix of M, then

$$\det(M)\xi_i = 0, \ i = 1,...,n.$$

This shows that $\det(M)B' = 0$. But $1_B = 1_{B'} \in B'$ (by our convention for subrings). Hence $\det(M) = 0$, and consequently b is a zero of some monic polynomial over R'. □

1.4. Corollary Let B be an R-algebra. If B is a finitely generated R-module, then B is integral over A. (One may compare this with a finite dimensional field extension.)

□

Example (iv) Let $d \in \mathbb{Z}$ be square-free. By previous Example (ii), $\mathbb{Z} \subset \mathbb{Z}[\sqrt{d}]$ is a module-finite extension and hence an integral extension.

(v) Let K be a field and $B = K[x]$ the polynomial ring in one variable x over K. Then for $n \geq 1$, $K[x^n] \subseteq K[x]$ is an integral extension. Indeed, one may check that $B = K[x^n] + K[x^n]x + K[x^n]x^2 + \cdots + K[x^n]x^{n-1}$, that is, B is module-finite over $K[x^n]$.

(vi) Let R be a UFD and K its field of fractions. Then any $y \in K - R$ is not integral over R. To see this, let $y = \frac{a}{b} \notin R$, where $\gcd(a,b) = d \in U(R)$. If $y^n + \lambda_{n-1}y^{n-1} + \cdots + \lambda_1 y + \lambda_0 = 0$, $\lambda_i \in R$, then, multiplying by b^n, we have

$$a^n = -b\left(\lambda_{n-1}a^{n-1} + \lambda_{n-2}ba^{n-2} + \cdots + \lambda_1 b^{n-2}a + b^{n-1}\lambda_0\right) \in R.$$

This shows that if a is a unit in R, then b is a unit in R and hence $y \in R$. If a is not a unit in R, then b must be a unit in R and $y = ab^{-1} \in R$; otherwise a and b would have a nonunit common divisor.

(vii) Let $K[x, y]$ be the polynomial ring in x, y over a field K. Consider the ring $R = K[x, y]/\langle y - f(x) \rangle$, where f is a monic polynomial with $\deg f \geq 1$. Then, $\overline{x}$ is integral over $K[\overline{y}]$ and R is module-finite over $K[\overline{y}]$.

(viii) Let $K[x, y]$ be the polynomial ring in x, y over a field K. Then, though $\overline{xy} - 1 = 0$ in the ring $R = K[x, y]/\langle yx - 1 \rangle$, $\overline{x}$ is not integral over $K[\overline{y}]$. Indeed, by Chapter 2 (section 3, Example (ii)), with $S = \{1, y, y^2, \ldots\}$,

$$K[y]_S = K[y][1/y] \cong \frac{K[y][x]}{\langle yx - 1 \rangle},$$

while $K[y][1/y]$ is not a finitely generated $K[y]$-module by Chapter 1 (section 7, exercise 3).

1.5. Proposition Let B be an R-algebra defined by the ring homomorphism $\varphi \colon R \to B$. Write $R' = \varphi(R)$. The following hold:
(i) The subset

$$\overline{R} = \left\{ b \in B \mid b \text{ is integral over } R \right\}$$

is an R-subalgebra of B.
(ii) If $b \in B$ is integral over $\overline{R}$, then $b \in \overline{R}$. Hence $\overline{\overline{R}} = \overline{R}$.

Proof (i) Let $b_1, b_2 \in \overline{R}$. We show that $b_1 \pm b_2$, $b_1 b_2 \in \overline{A}$. Consider the ring extension

$$R'[b_1] \subseteq R'[b_1][b_2] = R'[b_1, b_2].$$

By Theorem 1.3 (ii),

$$R'[b_1, b_2] = \sum_{i=1}^{m} R'[b_1] u_i, \ u_i \in R'[b_1, b_2],$$

$$R'[b_1] = \sum_{j=1}^{n} R' v_j, \ v_j \in R'[b_1].$$

Thus,

$$R'[b_1, b_2] = \sum_{i=1}^{m} \left(\sum_{j=1}^{n} R' v_j \right) u_i,$$

and it follows from Theorem 1.3(iii) that $b_1 \pm b_2$, $b_1 b_2 \in \overline{R}$.
(ii) Let $b \in B$ be such that

$$b^n + \lambda_{n-1} b^{n-1} + \cdots + \lambda_1 b + \lambda_0 = 0, \ \lambda_i \in \overline{R}.$$

Then, inductively we derive that

$$b^{n+i} \in \sum_{j=0}^{n-1} R'[\lambda_0, \lambda_1, ..., \lambda_{n-1}] b^j, \ i \in \mathbb{N}.$$

Hence $R'[b] \subseteq \sum_{j=0}^{n-1} R'[\lambda_0, \lambda_1, ..., \lambda_{n-1}] b^j$. Note that the λ_i are integral over R. A similar argument as for part (i) shows that $R'[\lambda_0, \lambda_1, ..., \lambda_{n-1}]$ is contained in an R-subalgebra of B which is also a finitely generated R'-module. By Theorem 1.3(iii), $b \in \overline{R}$. □

1.6. Definition The subalgebra $\overline{R}$ obtained in Proposition 1.5 is called the *integral closure* of R in B.
If $R \subseteq B$ and $R = \overline{R} \subseteq B$, then we say that R is *integrally closed* in B. (Compare this with the algebraic closure in a field extension defined in Chapter 1 (section 3, exercise 3).)

Combined with the forthcoming Noether normalization of section 2, the next proposition will play an important role in algebraic geometry (see the proof of Chapter 5 (Theorems 5.7 and 6.3)).

1.7. Proposition Let $R \subseteq B$ be a module-finite extension and $M \in$ m-Spec$R$. Then $MB \neq B$ and there exists a maximal ideal $Q \in$ m-Spec$B$ such that $Q \cap R = M$. (A generalization of this result is given in exercise 8 below.)

Proof Suppose $B = \sum_{i=1}^{s} R\xi_i$, $\xi_i \in B$. If $B = MB = \sum_{i=1}^{s} M\xi_i$, then $\xi_i = \sum_{i=1}^{s} r_{ij}\xi_i$, $r_{ij} \in M$, $i = 1, ..., s$. Arguing as in the proof ((iii) $\Rightarrow$ (i)) of Theorem 1.3, there would be $1 \in M$. Hence $MB \neq B$. Let Q be a maximal ideal of B that contains MB (Chapter 2 Proposition 1.5). Note that $R \cap Q \neq R$ (otherwise $1 \in Q$). Thus, $M \subset R \cap Q$ implies $M = R \cap Q$. □

Finally, let us see how integral extension is closely related to field extensions (see the proof of later Theorem 2.5, Chapter 5 (Theorems 1.5 and 5.3) for applications).

1.8. Theorem Let $R \subseteq B$ be an integral extension, where R and B are domains. Then R is a field if and only if B is a field.

Proof Suppose R is a field. If $0 \neq b \in B$ then

$$b^n + \lambda_{n-1}b^n + \cdots + \lambda_1 b + \lambda_0 = 0, \ \lambda_i \in R, \ n \geq 1.$$

Assume that n is the smallest degree. Then $\lambda_0 \neq 0$ and

$$b^{-1} = -\lambda_0^{-1}\left(b^{n-1} + \lambda_{n-1}b^{n-2} + \cdots + \lambda_2 b + \lambda_1\right) \in B.$$

Conversely, if B is a field and $0 \neq a \in R$, then $a^{-1} \in B$ and hence a^{-1} is integral over R. It follows that there is a relation of the form

$$(a^{-1})^n + \lambda_{n-1}(a^{-1})^{n-1} + \cdots + \lambda_1 a^{-1} + \lambda_0 = 0, \ \lambda_i \in R.$$

Consequently, $a^{-1} = -\lambda_{n-1} - \lambda_{n-2}a - \cdots - \lambda_0 a^{n-1}$, i.e., $a^{-1} \in R$.

Exercises

1. Let $R = \mathbb{Z}[\sqrt{p_i} \mid p_i$'s are distinct prime numbers, $i \geq 1]$. Show that R is neither module-finite nor finitely generated as an algebra over $\mathbb{Z}$. (See also previous Example (iii) and Chapter 1 (section 7, exercise 3).)

2. By previous Example (v), $K[x^2] \subset K[x]$ is an integral extension. For any $f \in K[x]$, find a monic polynomial $g(t) \in k[x^2][t]$ such that $g(f) = 0$.
3. Let $A \subseteq B \subseteq C$ be ring extensions. Suppose that C is module-finite over B and B is module-finite over A. Show that C is module-finite over A.
4. Let $R \subseteq B$ be a ring extension. Show that if $b_1, ..., b_m \in B$ are integral over R then $R[b_1, ..., b_m]$ is module-finite over R. Hence, B is module-finite over R if and only if B is finitely generated as an R-algebra and integral over R.
5. Let $A \subseteq B \subseteq C$ be ring extensions. Suppose that C is integral over B and B is integral over A. Show that C is integral over A.
6. Let $R \subseteq B$ be an integral extension.
   (a) For $u \in R$, show that $u \in U(R)$ if and only if $u \in U(B)$.
   (b) Let H be a proper ideal of B and $h = H \cap R$. Show that (up to a ring monomorphism) $R/h \subset B/H$ is an integral extension.
7. Let $R \subseteq B$ be an integral extension, and let S be a multiplicative set of R, $0 \notin S$. Then clearly S is also a multiplicative set in B.
   (a) Show that $R_S \subseteq B_S$, and that B_S is integral over R_S. (Hint: Use Chapter 2 Proposition 4.3 and Chapter 2 (section 3, exercise 2).)
   (b) Show that if B is module finite over R then B_S is module finite over R_S. (Hint: Use Chapter 2 (section 4, exercise 6).)
8. (Lying over theorem) Let $R \subseteq B$ be an integral extension. Show that if $P \in \mathrm{Spec} R$ then there is some $Q \in \mathrm{Spec} B$ such that $Q \cap R = P$. (Hint: First, by arguing as in the proof of Proposition 1.7, show that the assertion is true for every $M \in \mathrm{m\text{-}Spec} R$. Then, for each $P \in \mathrm{Spec} R$, use exercise 7 above, Chapter 2 (Corollary 3.5 and Proposition 4.3) in order to pass to the case $R_P \subseteq B_P$.)

2. Noether Normalization

Let $R = K[a_1, ..., a_n]$ be a finitely generated K-algebra over the field K. Then by Chapter 1 section 0, $R \cong K[x_1, ..., x_n]/I$, where I is an ideal of the polynomial K-algebra $K[x_1, ..., x_n]$. Hence R is Noetherian by Chapter 1 (section 1, exercise 2). In this section we show how to build R by a module-finite extension over some polynomial K-algebra.

By the definition of a polynomial ring we know that $\sum \lambda_{\alpha(j)} x_1^{\alpha_{j1}} \cdots x_n^{\alpha_{jn}}$

$= 0$ in $K[x_1, ..., x_n]$ if and only if all $\lambda_{\alpha(j)} = 0$. This property has the following generalization to an arbitrary K-algebra.

2.1. Definition Let R be a K-algebra over the field K, and $r_1, ..., r_m \in R$. If there is some polynomial $g \in K[x_1, ..., x_m]$ such that $g(r_1, ..., r_m) = 0$, then we say that $r_1, ..., r_m$ are *algebraically dependent* over the field K; otherwise, $r_1, ..., r_m$ are said to be *algebraically independent* over the field K.

Clearly, if $K \subseteq L$ is a field extension and $\vartheta \in L$ is a transcendental element over K (Chapter 1 section 3), then ϑ is algebraically independent over K and $K[\vartheta]$ is isomorphic to the polynomial algebra $K[x]$. More generally, if R is a K-algebra over the field K and $r_1, ..., r_m \in R$ are algebraically independent over K, then the subalgebra $K[r_1, ..., r_m]$ of R is isomorphic to the polynomial K-algebra $K[x_1, ..., x_m]$.

To reach the main result of this section, the following technical preliminary is required.

2.2. Lemma Let $\alpha(j) = (\alpha_{j_1}, ..., \alpha_{j_n})$, $\alpha(k) = (\alpha_{k_1}, ..., \alpha_{k_n}) \in \mathbb{N}^n$. Suppose $\sum_{i=1}^{n} \alpha_{j_i} = \sum_{i=1}^{n} \alpha_{k_i} = m$, but lexicographically (see Chapter 1 section 4) $\alpha(k) \prec_{lex} \alpha(j)$, that is, there is some $s \leq n$ such that

$$\alpha_{k_1} = \alpha_{j_1}, ..., \alpha_{k_{s-1}} = \alpha_{j_{s-1}}, \text{ while } \alpha_{k_s} < \alpha_{j_s}.$$

Then $\sum_{i=1}^{n} m^{n-i} \alpha_{k_i} = B < A = \sum_{i=1}^{n} m^{n-i} \alpha_{j_i}$.

Proof Without loss of generality we may assume $\alpha_{k_1} < \alpha_{j_1}$. Then

$$B = m^{n-1} \alpha_{k_1} + m^{n-2} \alpha_{k_2} + \cdots + m \alpha_{k_{n-1}} + \alpha_{k_n}$$

$$\leq m^{n-1}(\alpha_{j_1} - 1) + m^{n-2} \alpha_{k_2} + \sum_{i=3}^{n} m^{n-i} \alpha_{k_i}$$

$$= m^{n-1} \alpha_{j_1} + m^{n-2}(\alpha_{k_2} - m) + \sum_{i=3}^{n} m^{n-i} \alpha_{k_i}$$

$$= m^{n-1} \alpha_{j_1} + m^{n-2} \left(\alpha_{k_2} - \sum_{i=1}^{n} \alpha_{k_i} \right) + \sum_{i=3}^{n} m^{n-i} \alpha_{k_i}$$

$$= m^{n-1}\alpha_{j_1} - m^{n-2}\alpha_{k_1} - \sum_{i=3}^{n} m^{n-2}\alpha_{k_i} + \sum_{i=3}^{n} m^{n-i}\alpha_{k_i}$$

$$< \sum_{i=1}^{n} m^{n-i}\alpha_{j_i}$$

$$= A.$$

□

Furthermore, let $f \in K[x_1, ..., x_n]$ be a polynomial of $\deg f = m \geq 1$, where K is a field. Then by Chapter 1 (section 7, exercise 8), f may be written uniquely as

$$f = F_m + F_{m-1} + \cdots + F_0,$$

where each F_i is a homogeneous polynomial of degree n_i, i.e.,

$$F_i = \sum_{\alpha_{j_1} + \cdots + \alpha_{j_n} = n_i} \lambda_{\alpha(j)} x_1^{\alpha_{j_1}} x_2^{\alpha_{j_2}} \cdots x_n^{\alpha_{j_n}}$$

with $\alpha(j) = (\alpha_{j_1}, ..., \alpha_{j_n}) \in \mathbb{N}^n$, $\lambda_{\alpha(j)} \in K$.

Let us call F_m the *leading homogeneous part* of f for convenience.

2.3. Proposition Let $R = K[a_1, ..., a_n]$ be a finitely generated K-algebra over the field K. Suppose that there is a nonzero polynomial $f \in K[x_1, ..., x_n]$ such that $f(a_1, ..., a_n) = 0$. Then there exist $b_1, ..., b_{n-1} \in R$ such that a_n is integral over $K[b_1, ..., b_{n-1}]$ and $R = K[b_1, ..., b_{n-1}][a_n]$.

Proof Write $f = F_m + F_{m-1} + \cdots + F_0$ as a sum of homogeneous polynomials, where the leading homogeneous part of f has the form

(1) $\quad F_m = \sum_{j=1}^{q} \lambda_{\alpha(j)} x_1^{\alpha_{j_1}} x_2^{\alpha_{j_2}} \cdots x_n^{\alpha_{j_n}}$ with $\alpha(j) = (\alpha_{j_1}, ..., \alpha_{j_n})$, $\lambda_{\alpha(j)} \neq 0$.

Since $\preceq_{lex}$ is a total ordering on $\mathbb{N}^n$, we may also assume that

(2) $\quad\quad\quad\quad \alpha(j) \prec_{lex} \alpha(1),\ 1 < j \leq q.$

Now set
$$\mu_i = m^{n-i}, \quad i = 1, 2, ..., n-1,$$
$$x_i^* = x_i - x_n^{\mu_i},$$

and define

(3) $\quad H = H\left(x_1^*, ..., x_{n-1}^*, x_n\right) = f\left(x_1^* + x_n^{\mu_1}, ..., x_{n-1}^* + x_n^{\mu_{n-1}}, x_n\right).$

Then it follows from (1) + (2) and Lemma 2.2 that

$$H = F_m\left(x_1^* + x_n^{\mu_1}, ..., x_{n-1}^* + x_n^{\mu_{n-1}}, x_n\right) + \cdots + F_0$$

$$= \left(\sum_{j=1}^{q} \lambda_{\alpha(j)} \left(x_1^* + x_n^{\mu_1}\right)^{\alpha_{j_1}} \cdots \left(x_{n-1}^* + x_n^{\mu_{n-1}}\right)^{\alpha_{j_{n-1}}} x_n^{\alpha_{j_n}}\right) + \cdots + F_0,$$

in which the highest degree term in x_n is given by

$$F_m\left(x_1^* + x_n^{\mu_1}, ..., x_{n-1}^* + x_n^{\mu_{n-1}}, x_n\right),$$

that is,

$$\lambda_{\alpha(1)} x_n^{\sum \mu_i \alpha_{1_i}}.$$

Setting $b_k = a_k - a_n^{\mu_k}$, $k = 1, ..., n-1$, we see from (3) above that

$$H(b_1, ..., b_{n-1}, a_n) = f(a_1, ..., a_n) = 0.$$

Note that $0 \neq \lambda_{\alpha(1)} \in K$. We conclude that a_n is a zero of the monic polynomial $\lambda_{\alpha(1)}^{-1} H(b_1, ..., b_{n-1}, t) \in K[b_1, ..., b_{n-1}][t]$.

Finally, it is clear that $R = K[b_1, ..., b_{n-1}][a_n]$, and thus, the proof is completed. $\square$

2.4. Theorem (Noether normalization) Let $R = K[a_1, ..., a_n]$ be a finitely generated K-algebra over a field K. Then there exist elements $z_1, ..., z_d \in R$, $0 \leq d \leq n$, such that
(i) $z_1, ..., z_d$ are algebraically independent over K;
(ii) R is module-finite over the polynomial K-algebra $B = K[z_1, ..., z_d]$.

Proof We prove by induction on the number n of generators of R.

For $n = 1$, the assertion follows from (section 1, exercise 4). So, For $n \geq 1$, if $a_1, ..., a_n$ are algebraically independent over K, then the assertion is trivially true. Suppose that $a_1, ..., a_n$ are algebraically dependent over K, and that $f \in K[x_1, ..., x_n]$ is such that $f(a_1, ..., a_n) = 0$. Then, by

Proposition 2.3, there exist $b_1, ..., b_{n-1} \in R$ such that a_n is integral over $B = K[b_1, ..., b_{n-1}]$ and $R = B[a_n]$. Applying the inductive hypothesis to B, we may find $z_1, ..., z_d \in B$, that are algebraically independent over K, such that B is module-finite over $A = K[z_1, ..., z_d]$. Since a_n is integral over B, it follows that $B[a_n]$ is module-finite over B. If we look at each step of the tower $A \subset B \subset B[a_n] = R$, then by (section 1, exercise 3), R is module finite over A, which completes the proof. □

Theorems 1.8 and 2.4 enable us to derive the following wonderful result that is crucial in dealing with the Nullstellensatz in algebraic geometry (Chapter 5 section 1).

2.5. Theorem (Zariski) Let $K \subseteq L$ be a field extension. If $L = K[\alpha_1, ..., \alpha_m]$ is a finitely generated K-algebra, then L is module-finite over K, i.e., $[L : K] < \infty$, and hence L is algebraic over K.

Proof By the Noether normalization, there are $z_1, ..., z_d \in L$, which are algebraically independent over K, such that

$$L = \sum_{j=1}^{n} K[z_1, ..., z_d]\xi_j, \ \xi_j \in L.$$

Thus, $K[z_1, ..., z_d] \subseteq L$ is an integral extension. But since L is a field, it follows from Theorem 1.8 that $K[z_1, ..., z_d]$ is a field. This shows that $d = 0$, and consequently L is module-finite over K.

Exercises

1. Given distinct $\alpha(1), \alpha(2), ..., \alpha(m) \in \mathbb{N}^n$ with $\alpha(j) = (\alpha_{j_1}, \alpha_{j_2}, ..., \alpha_{j_n})$, show that there are nonnegative integers $\mu_1, ..., \mu_{n-1}, 1 = \mu_n$ such that

$$\sum \mu_i \alpha_{j_i} \neq \sum \mu_i \alpha_{h_i} \text{ whenever } \alpha(j) \neq \alpha(h).$$

(Hint: Use induction on n. If $n = 1$, it is trivially true. Suppose for $\alpha'(1), ..., \alpha'(m) \in \mathbb{N}^{n-1}$ with $\alpha'(j) = (\alpha'_{j_2}, ..., \alpha'_{j_m})$ we can choose $\mu_2, ..., \mu_{n-1}, \mu_n = 1$ so that the conclusion is true. Now the choice

$$\mu_1 > \max\left\{\sum_{i=2}^{n} \mu_i \alpha_{j_i} \ \middle| \ j = 1, ..., m\right\}$$

completes the induction process.)

2. Let $R = K[x,y]/\langle y^2 - x^3\rangle$, where $K[x,y]$ is the polynomial ring in x, y over a field K. Show that $\bar{x}$ is algebraically independent over K, $\bar{y}$ is integral over $K[\bar{x}]$, and $R = k[\bar{x},\bar{y}]$ is module-finite over $K[\bar{x}]$.
3. Let $K[x,y,z]$ be the polynomial ring in x,y,z over a field K, $R = K[x,y,z]/\langle xz, y^2z\rangle$. Write $u = \bar{x} + \bar{z}$, $v = \bar{y}$. Show that u and v are algebraically independent over K and that R is module-finite over the subring $K[u,v]$.
4. Let $K[x,y]$ be the polynomial ring in x,y over a field K. By (section 1, Example (viii)), in the ring $R = K[x,y]/\langle yx - 1\rangle$, $\bar{x}$ is not integral over $K[\bar{y}]$, and similarly $\bar{y}$ is not integral over $K[\bar{x}]$. Use Noether normalization to find some $z \in R$ that is algebraically independent over K, such that R is module-finite over $K[z]$.

3. Normal Domains and Normalization

In this section we introduce the notion of a normal domain, and establish the module-finite property of the normalization (integral closure) of a finitely generated domain.

3.1. Definition Let R be a domain, K its field of fractions, and $\overline{R}$ its integral closure in K (Definition 1.6).
(i) If R is integrally closed in K, i.e., $R = \overline{R} \subset K$, then we say that R is a *normal domain*.
(ii) If $R = F[a_1,...,a_n]$ is a finitely generated F-algebra over some field F, then $\overline{R}$ is called the *normalization* of R (this name will be qualified by the proof of Corollary 3.3 below and by the behavior of $\overline{R}$ in Chapter 5 section 5).

Example (i) Let R be a domain and K its field of fractions. If $\overline{R}$ is the integral closure of R in K, then, since the field of fractions of $\overline{R}$ is also K (why?) and $\overline{\overline{R}} = \overline{R} \subset K$, by Definition 3.1, $\overline{R}$ is a normal domain.

By (section 1, Example (vi)), any UFD is normal. In particular, any DVR is normal.

(ii) Let $K = \mathbb{Q}(\vartheta)$ be a number field and $\mathcal{A}_K$ its ring of algebraic integers. Then $\mathcal{A}_K$ is normal and Noetherian but not necessarily a UFD (see Chapter 4 for details).

(iii) Let $K[x,y]$ be the polynomial ring in variables x,y over a field K, and let $R = K[x,y]/\langle y^2 - x^3 \rangle$ (see section 2, exercise 2). Then $\overline{y}^2 = \overline{x}^3$ in R. Thus, any element of R has the form

$$\overline{\sum \lambda_{nm} x^n y^m} = \overline{\sum \lambda_{nm} x^n y^{2m'+r_m}}, \text{ (where } 0 \leq r_m \leq 1)$$

$$= \sum \lambda_{nm} \overline{x}^{n+3m'} \overline{y}^{r_m}$$

$$= \sum \lambda_{n_1 m_1} \overline{x}^{n_1 + 3m_1} \overline{y} + \sum \lambda_{n_2 m_2} \overline{x}^{n_2 + 3m_2}.$$

Let $K[t]$ be the polynomial ring in variable t over K. Then the ring homomorphism $K[x,y] \to K[t^2, t^3]$ with $x \mapsto t^2$ and $y \mapsto t^3$ induces a ring homomorphism (see Chapter 1 section 0):

$$\begin{aligned} \varphi: R &\longrightarrow K[t^2, t^3] \\ \overline{x} &\mapsto t^2 \\ \overline{y} &\mapsto t^3 \end{aligned}$$

By the remark made above, it is not difficult to see that φ is an isomorphism. Note that $\frac{t^3}{t^2} = t$ in the field of fractions $K(t)$ of $K[t]$. Identifying R with the subring $K[t^2, t^3]$ of $K[t]$, R and $K[t]$ have the same field of fractions $K(t)$, i.e., we have the relation

$$R \subset K[t] \subset K(t).$$

Clearly, t is integral over R but is not contained in R. Hence, R is not normal. But since $K[t]$ is a UFD, the normalization $\overline{R}$ of R is $K[t]$ by Example (i) above. A geometric explanation of this example will be given in Chapter 5 section 5.

(iv) Let $K[x,y,z]$ be the polynomial ring in x,y,z over a field K, $R = K[x,y,z]/\langle zy^2 - x^2 \rangle$. If $K[u,v]$ is the polynomial ring in u,v over K, then the ring homomorphism $K[x,y,z] \to K[uv, u, v^2]$ with $x \mapsto uv$, $y \mapsto u$, and $z \mapsto v^2$ induces a ring homomorphism

$$\begin{aligned} \psi: R &\longrightarrow K[uv, u, v^2] \\ \overline{x} &\mapsto uv \\ \overline{y} &\mapsto u \\ \overline{z} &\mapsto v^2 \end{aligned}$$

A similar argumentation as in Example (iii) above shows that ψ is an isomorphism, and the normalization $\overline{R}$ of R is $K[u,v] \neq K[uv, u, v^2]$.

By (section 1, exercise 4), the normalization $\overline{R}$ of R obtained in Example (iii), respectively in Example (iv) above, is module-finite over R (see also other examples provided by exercises of the current section). Below we will see that this is by no means accidental.

3.2. Theorem Let R be a normal domain and K its field of fractions. If $K \subset L = K(\vartheta)$ is an n-dimensional separable field extension, then the integral closure of R in L, denoted $\overline{R}$, is contained in a finitely generated R-submodule of L. If R is Noetherian then $\overline{R}$ is module-finite over R.

Proof Choose elements $\alpha_1, ..., \alpha_n \in \overline{R}$ to form a K-basis for L (why is this possible?). Then $\sum_{i=1}^{n} R\alpha_i \subseteq \overline{R}$. Consider the trace function $T_{L/K}$ on L as defined in Chapter 1 section 5. By Chapter 1 (Theorems 5.5 and 5.7), $T_{L/K}(uv)$ is a nondegenerate bilinear form and there exists a dual K-basis $\{\beta_1, ..., \beta_n\}$ for L such that

(1) $$T_{L/K}(\alpha_i \beta_j) = \begin{cases} 0, & \text{if } i \neq j, \\ 1, & \text{if } i = j. \end{cases}$$

We claim that $\overline{R} \subseteq \sum_{i=1}^{n} R\beta_i$. To see this, let $\sigma_1, ..., \sigma_n$ be all distinct K-linear monomorphisms from L to a splitting field of $p_\theta(x)$ (the minimal polynomial of θ over K). Then for $u \in \overline{R} \subseteq L$, we have $\sigma_i(u) \in \overline{R}$, $i = 1, ..., n$. But $\sum_{i=1}^{n} \sigma_i(u) = T_{L/K}(u) \in K$, it follows that $T_{L/K}(u) \in R$ because R is normal. Now, for $\gamma \in \overline{R}$, $\gamma = \sum_{i=1}^{n} \lambda_i \beta_i$ with $\lambda_i \in K$, we have $T_{L/K}(\gamma \alpha_j) \in R$ and by the above (1),

$$\lambda_j = \sum_{i=1}^{n} \lambda_i T_{L/K}(\beta_i \alpha_j) = T_{L/K}(\gamma \alpha_j) \in R, \ j = 1, ..., n.$$

This shows that $\overline{R} \subseteq \sum_{i=1}^{n} R\beta_i$, as claimed.
The last assertion follows from Chapter 1 Theorem 7.6. □

3.3. Corollary Let $R = F[a_1, ..., a_m]$ be a finitely generated F-algebra over a field F. If R is a domain and K is its field of fractions, then the normalization $\overline{R}$ of R in K is module-finite over R and hence a Noetherian ring.

Proof By the Noether normalization theorem, there are algebraically independent $z_1, ..., z_d \in R$ over F such that

$$F \subseteq F[z_1, ..., z_d] \subseteq R = \sum_{j=1}^{n} F[z_1, ..., z_d]\xi_j, \ \xi_j \in R.$$

Assume that K is a separable extension of $F(z_1, ..., z_d)$ that is the field of fractions of the polynomial ring $F[z_1, ..., z_d]$. Then since $\overline{R}$ is also the integral closure of $F[z_1, ..., z_d]$ in K, it follows from Theorem 3.2 that $\overline{R}$ is module-finite over R (note that R is Noetherian). □

Applications of Theorem 3.2 and Corollary 3.3 are given in Chapter 4 and Chapter 5.

Remark In the proof of Corollary 3.3 we argued by assuming the separability of K over $F(z_1, ..., z_d)$. If the ground field F is of characteristic 0, the separability of K is naturally guaranteed. In general case, the feasibility of this assumption is proved in detail by Zariski and Samuel in Theorem 9 (p. 267) of *Commutative Algebra*, Vol. I (New York: Springer-Verlag, 1958). We do not quote Zariski-Samuel's detailed argumentation here because it involves more deeper field theory that is not included in Chapter 1. So the reader may accept Corollary 3.3 and use it without any doubt.

Exercises

1. Let $R = K[x, y]/\langle y - f(x)\rangle$, where $K[x, y]$ is the polynomial ring in x, y over a field K and $\deg f(x) \geq 1$. Is R a normal domain?
2. Let $R = K[x, y]/\langle y^2 - x^2 - x^3\rangle$ where $K[x, y]$ is the polynomial ring in x and y over a field K. Prove that the normalization of R is isomorphic to the polynomial ring $K[t]$. (See also Chapter 5 (section 7, Example (iv)).)
3. Let $R = K[x, y, z]/\langle y - x^2, z - x^3\rangle$, where $K[x, y, z]$ is the polynomial ring in x, y, z over a field K. Show that R is a normal domain. (Hint: $R \cong K[t]$ with $x \mapsto t$, $y \mapsto t^2$, and $z \mapsto t^3$.)

4. Normal Domains and DVRs

This section is devoted to a local study of normal domains in terms of DVRs.

First, a typical example of "global concern versus local solutions".

4.1. Theorem Let R be a domain and K its field of fractions. The following statements are equivalent.
(i) R is normal.
(ii) R_S is normal for any multiplicative set S of R.
(iii) R_P is normal for all $P \in \operatorname{Spec} R$.
(iv) R_M is normal for all $M \in \text{m-Spec} R$.

Proof (i) $\Rightarrow$ (ii) let $x \in K$ be integral over R_S. Then
$$x^n + \lambda_{n-1} x^{n-1} + \cdots + \lambda_1 x + \lambda_0 = 0, \quad \lambda_i = \frac{r_i}{s_i}, \; r_i \in R, \; s_i \in S.$$

Multiplying the relation by $(s_0 s_1 \cdots s_{n-1})^n$, we see that $(s_0 s_1 \cdots s_{n-1})x$ is integral over R. By the fact that R is normal, $(s_0 s_1 \cdots s_{n-1})x = y \in R$, i.e.,
$$x = \frac{y}{s_0 s_1 \cdots s_{n-1}} \in R_S.$$

(ii) $\Rightarrow$ (iii) $\Rightarrow$ (iv) is obvious.
(iv) $\Rightarrow$ (i) let $x \in K$ be integral over R. Then
$$x^n + \lambda_{n-1} x^{n-1} + \cdots + \lambda_1 x + \lambda_0 = 0$$
holds over both R and R_M for each $M \in \text{m-Spec} R$ (note that $R \subseteq R_M$). Since R_M is normal, $x \in \cap R_M = R$ by Chapter 2 Theorem 3.8. $\square$

Comparing with Chapter 2 Proposition 2.8, our goal is to show that the equality $R = \cap R_P$ holds for a normal and Noetherian domain R, where P runs over all minimal nonzero primes of R and every R_P is a DVR, though a normal and Noetherian domain is not necessarily a UFD (section 3, Example (ii)). To this end, we need the following preparation.

Let R be *any* ring, M an R-module, and $x \in M$. Then the set of annihilators of x in R, denoted
$$\operatorname{Ann}_R x = \left\{ a \in R \;\middle|\; ax = 0 \right\}$$

is an ideal of R. Indeed, $\mathrm{Ann}_R x$ coincides with the kernel of the R-module homomorphism

$$\rho_x : R \longrightarrow M$$
$$a \longmapsto ax$$

Hence $R/\mathrm{Ker}\rho_x = R/\mathrm{Ann}_R x$ is isomorphic to an R-submodule of M. Note that if $x \neq 0$ then $\mathrm{Ann}_R x \neq R$, for $1_R \in R$ and all modules considered in this book are unitary.

If $\mathrm{Ann}_R x = P \in \mathrm{Spec}R$, we say that P is an *associated prime* of M. Set

$$\mathrm{Ass}_R M = \Big\{ P \in \mathrm{Spec}R \ \Big| \ P \text{ an associated prime of } M \Big\}.$$

Example (i) For any $P \in \mathrm{Spec}R$, $\mathrm{Ass}_R(R/P) = \{P\}$.

(ii) Let R be a PID, and let $M = R/\langle a \rangle$, where $a = p_1^n p_2^m$ for two primes $p_1 \neq p_2$. Then $\bar{u} = \overline{p_1^{n-1} p_2^m} \in M$ has $\mathrm{Ann}_R \bar{u} = \langle p_1 \rangle$, and similarly $\bar{v} = \overline{p_1^n p_2^{m-1}} \in M$ has $\mathrm{Ann}_R \bar{v} = \langle p_2 \rangle$. Moreover, if $w \in M$ with $\mathrm{Ann}_R w = \langle r \rangle \in \mathrm{Spec}R$, then, it follows from the first isomorphism theorem that $r | a$. Hence $\mathrm{Ass}_R M = \{\langle p_1 \rangle, \langle p_2 \rangle\}$.

4.2. Proposition Let R be a ring and M an R-module. The following hold:
(i) Any maximal element (with respect to the inclusion relation) in the set

$$\Omega = \Big\{ \mathrm{Ann}_R x \ \Big| \ 0 \neq x \in M \Big\}$$

is prime and hence belongs to $\mathrm{Ass}_R M$.
(ii) If R is Noetherian and $M \neq \{0\}$, then $\mathrm{Ass}_R M \neq \emptyset$.

Proof (i) Let $x \in M$ be chosen such that $\mathrm{Ann}_R x$ is maximal in Ω. If $a, b \in R$, $ab \in \mathrm{Ann}_R x$, then $abx = 0$. It follows that if $bx = 0$ then $b \in \mathrm{Ann}_R x$; if $bx \neq 0$, then since $bx \in M$ and $\mathrm{Ann}_R x \subseteq \mathrm{Ann}_R(bx)$, we have $\mathrm{Ann}_R(bx) = \mathrm{Ann}_R x$ by the choice of $\mathrm{Ann}_R x$. Therefore, $abx = 0$ yields $a \in \mathrm{Ann}_R x$.
(ii) If R is Noetherian and $M \neq \{0\}$, then

$$\Omega = \Big\{ \mathrm{Ann}_R x \ \Big| \ 0 \neq x \in M \Big\} \neq \emptyset$$

and has a maximal element. Hence $\mathrm{Ass}_R M \neq \emptyset$ by (i). $\square$

The next proposition provides the key result in the local study of a normal Noetherian domain.

4.3. Proposition Let R be a local domain with the unique maximal ideal $\mathbf{m}$. Assume that R is normal and Noetherian. If $\mathbf{m} \in \mathrm{Ass}_R(R/\langle a \rangle)$ for some $0 \neq a \in R$, then R is a DVR, and hence $\mathbf{m}$ is a minimal nonzero prime ideal.

Proof Let $u \in R - \langle a \rangle$ such that $\mathbf{m} u \subset \langle a \rangle$. Then $\mathbf{m}\frac{u}{a} \subseteq R$. If $\mathbf{m}\frac{u}{a} \subseteq \mathbf{m}$, then since $\mathbf{m}$ is a finitely generated R-module, a similar demonstration as in the proof of Theorem 1.3 ((iii) $\Rightarrow$ (i)) shows that $\frac{u}{a}$ is integral over R (note that now R is a domain) and hence in R. Thus, $u = ar \in \langle a \rangle$ for some $r \in R$. This contradicts the choice of u. Therefore, we must have $\mathbf{m}\frac{u}{a} = R$ because $\mathbf{m}\frac{u}{a}$ is an ideal of R.

Let $1 = v\frac{u}{a}$ for some $v \in \mathbf{m}$. Then, for any $y \in \mathbf{m}$, $y\frac{u}{a} = \frac{y}{v} \in R$ implies $y \in \langle v \rangle$. This shows that $\mathbf{m} = \langle v \rangle$. It follows from Chapter 2 Theorem 2.10 that R is a DVR and $\mathbf{m}$ is a minimal nonzero prime ideal. □

4.4. Corollary Let R be a normal and Noetherian domain and $P \in \mathrm{Ass}_R(R/\langle a \rangle)$ for some $0 \neq a \in R$. Then P is a minimal nonzero prime ideal of R.

Proof Consider the localization $(R/\langle a \rangle)_P$ of $R/\langle a \rangle$ at $R - P$. Then

$$(R/\langle a \rangle)_P = R_P/\langle a \rangle_P = R_P/aR_P$$

by Chapter 2 Corollary 4.4. Now $P \in \mathrm{Ass}_R(R/\langle a \rangle)$ implies $PR_P \in \mathrm{Ass}_{R_P}(R_P/aR_P)$ by later Exercise 4, and the assertion follows from Proposition 4.3 and Chapter 2 Proposition 3.6. □

Comparing with Chapter 2 (Theorems 2.6 and 2.10), we now have the following characterization of a DVR.

4.5. Theorem Let R be a domain. The following statements are equivalent.
(i) R is a DVR.
(ii) R is a normal and Noetherian local domain with $\mathrm{Spec} R = \{\{0\}, \mathbf{m}\}$ where $\mathbf{m}$ is the maximal ideal of R.

(Therefore, this theorem may also be viewed as the converse of Chapter 2 Theorem 2.6.)

Proof (i) ⇒ (ii) This follows from (section 3, Example (i)) and Chapter 2 Theorem 2.6.

(ii) ⇒ (i) By Chapter 2 Theorem 2.10, we need only to prove that (ii) implies **m** is a principal ideal.

By Nakayama's lemma, $\mathbf{m} \neq \mathbf{m}^2$. Let $0 \neq a \in \mathbf{m} - \mathbf{m}^2$. If $\langle a \rangle$ is a prime ideal then $\mathbf{m} = \langle a \rangle$ as desired. If $\langle a \rangle$ is not a prime ideal, then $\langle a \rangle \neq \mathbf{m}$ and there are $y, z \in R$, $y, z \notin \langle a \rangle$ but $yz \in \langle a \rangle$. This shows that $\mathrm{Ann}_R \bar{z} \neq \{0\}$, where $\bar{z}$ is the image of z in $M = R/\langle a \rangle$. By Proposition 4.2, $\mathrm{Ass}_R M \neq \{\{0\}\}$ and hence $\mathbf{m} \in \mathrm{Ass}_R M$. It follows from Proposition 4.3 (or its proof) that **m** is principal and R is a DVR. □

4.6. Corollary Let R be normal and Noetherian. If P is a minimal nonzero prime ideal of R (i.e., P is minimal in Spec$R$ with respect to the inclusion relation), then R_P is a DVR.

Proof R_P is normal by Theorem 4.1, and Noetherian by Chapter 2 Corollary 3.5. Since P is minimal in Spec$R$, it follows from Chapter 2 Proposition 3.6 that Spec$R_P = \{0, PR_P\}$. Hence R_P is a DVR by the theorem above. □

We are ready to prove the previously promised result.

4.7. Theorem Let R be a normal and Noetherian domain, K its field of fractions. Then

$$R = \bigcap R_P$$

where P runs over all minimal nonzero prime ideals of R. In particular, R is an intersection of DVRs.

Proof $R \subseteq \cap R_P$ is clear. To get the inclusion $\cap R_P \subseteq R$, a trick used in the proof of Chapter 2 Theorem 3.8 is employed again. Let $x = \frac{b}{c} \in K$, and consider the ideal of R given by

$$I_x = \left\{ a \in R \mid ax \in R \right\}.$$

Then $I_x = \{a \in R \mid ab \in \langle c \rangle\} = \mathrm{Ann}_R \bar{b}$, where $\bar{b}$ is the class of b in $R/\langle c \rangle$. If $\bar{b} = 0$, then $b \in \langle c \rangle$ and hence $x \in R$. Suppose $\bar{b} \neq 0$. Then

$$\Omega = \left\{ \mathrm{Ann}_R \bar{u} \mid 0 \neq \bar{u} \in R/\langle c \rangle \right\} \neq \emptyset.$$

Since R is Noetherian, I_x is contained in a maximal element P of Ω. By Proposition 4.2(i), $P \in \text{Ass}_R(R/\langle c \rangle)$. It follows from Corollary 4.4 that P is a minimal nonzero prime. However, note that $I_x \subseteq P$ implies $x \notin R_P$. This proves the following: $x \notin R$ implies $x \notin R_P$ for some minimal nonzero prime P.

Exercises

1. Let M be an R-module and N an R-submodule of M. Show that $\text{Ass}_R M \subseteq \text{Ass}_R N \cup \text{Ass}_R(M/N)$. Can you find an example to show that $\subsetneq$ holds?
2. Show that for any $n \geq 1$, there is a $\mathbb{Z}$-module M such that $\text{Ass}_\mathbb{Z} M$ contains exactly n prime ideals. (Hint: Consider $m = p_1^{\alpha_1} \cdots p_n^{\alpha_n}$ with distinct pimes p_i.)
3. Use Proposition 4.2(i) to show that if R is Noetherian and M an R-module, then
$$\bigcup_{0 \neq x \in M} \text{Ann}_R x = \bigcup_{P \in \text{Ass}_R M} P.$$
4. Let M be a nonzero R-module and $P \in \text{Ass}_R M$. If S is a multiplicative set of R with $P \cap S = \emptyset$, show that $PR_S \in \text{Ass}_{R_S} M_S$.

Chapter 4
The Ring $\mathcal{A}_K$ in $K = \mathbb{Q}(\vartheta)$

This chapter introduces the normal and Noetherian domain $\mathcal{A}_K$ that has its location among numbers as indicated by the following Hasse diagram (of the obvious partial order):

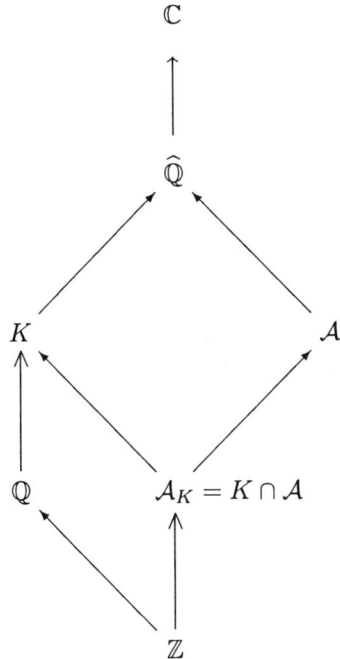

where

$$\widehat{\mathbb{Q}} = \left\{\alpha \in \mathbb{C} \mid f(\alpha) = 0 \text{ for some } f(x) \in \mathbb{Q}[x]\right\},$$

$$\mathcal{A} = \left\{\beta \in \mathbb{C} \mid g(\beta) = 0 \text{ for some monic } g(x) \in \mathbb{Z}[x]\right\},$$

$K \subset \mathbb{C}$ a subfield with $[K : \mathbb{Q}] < \infty$.

Consequently, $\widehat{\mathbb{Q}}$ is the algebraic closure of $\mathbb{Q}$ in $\mathbb{C}$ (Chapter 1 section 3, Exercise 3), $\mathcal{A}$ is the integral closure of $\mathbb{Z}$ in $\mathbb{C}$ (Chapter 3 Definition 1.6), and $\mathcal{A}_K$ is the integral closure of $\mathbb{Z}$ in K (Chapter 3 Definition 1.6). Very soon in section 1 below we will see that $\mathcal{A}_K$ is the integral closure of $\mathbb{Z}[\vartheta]$, where $K = \mathbb{Q}(\vartheta)$.

For any $\beta \in \mathcal{A}$, since $\mathcal{A}$ is integrally closed in $\mathbb{C}$, we have $\sqrt{\beta} \in \mathcal{A}$ (Chapter 3 Proposition 1.5(ii)). It follows that

$$\alpha = \alpha^{\frac{1}{2}} \cdot \alpha^{\frac{1}{2}} = \left(\alpha^{\frac{1}{4}} \cdot \alpha^{\frac{1}{4}}\right)\left(\alpha^{\frac{1}{4}} \cdot \alpha^{\frac{1}{4}}\right) = \cdots$$

holds for all $\alpha \in \mathcal{A}$. Note that not every $\alpha \in \mathcal{A}$ is a unit (for instance $\sqrt{5}$). We conclude that *factorization into irreducible elements in $\mathcal{A}$ is impossible*, and consequently, $\mathcal{A}$ is not Noetherian (Chapter 1 Proposition 2.5). Thus, $\mathcal{A}$ is usually not involved in practice.

Why $\mathcal{A}_K$? First and foremost, $\mathcal{A}_K$ is a normal and Noetherian domain (section 1, Theorem 1.2 below). So factorization into irreducible elements in $\mathcal{A}_K$ is always feasible (Chapter 1 Proposition 2.5), and in many cases $\mathcal{A}_K$ is a UFD (section 3). Perhaps a motive example to this topic should be the following oldest example.

Example Find all integer solution (x, y, z) of the equation

(1) $$x^2 + y^2 = z^2$$

where $\gcd(x, y, z) = 1$.

Solution In order to solve this problem, one way is to factorize the left of equation (1) in the ring of Gaussian integers $\mathbb{Z}[i] = \{a + bi \mid a, b \in \mathbb{Z}\}$ with $i = \sqrt{-1}$:

(2) $$(x + yi)(x - yi) = z^2.$$

Note that $\mathbb{Z}[i]$ is a UFD (see section 3). The assumption $(x, y, z) = 1$ entails that x and y cannot be both even or odd. So, z must be odd. Thus, $x + yi$ and $x - yi$ must be coprime, and a comparison of the prime factorizations on both sides of equation (2) yields $x + yi = u\alpha^2$, where u is a unit of $\mathbb{Z}[i]$ and $\alpha = c + di$ which is not a unit. Since $U(\mathbb{Z}[i]) = \{\pm 1, \pm i\}$ (see section 3), it follows that the desired solutions (x, y, z) are given by $x = \pm(c^2 - d^2)$, $y = \pm 2cd$, $z = \pm(c^2 + d^2)$.

In Chapter 5 (section 7, Example (iii)), another way to find the integer solutions of equation (1) will be given in terms of algebraic geometry.

In section 2 we will see that actually $\mathbb{Z}[i] = \mathbb{Q}(i) \cap \mathcal{A}$.

A generalization of last example is the famous **Fermat's last theorem** which states:

- For $n > 2$, the equation $x^n + y^n = z^n$ does not have nonzero integer solutions (x, y, z).

Using the above example, one may show that the theorem is true for $n = 4$ and hence (automatically) also for $n = 4k$. It is therefore sufficient to consider the case where n is an odd prime p, for if no solutions exist when $n = p$ then no solutions exist when $n = p\ell$. Now, for p, an odd prime,

$$y^p = z^p - x^p = (z - x)(z - \omega x)(z - \omega^2 x) \cdots (z - \omega^{p-1} x)$$

in $\mathbb{Z}[\omega]$, where ω is a primitive pth root of unity, i.e., $\omega = e^{2\pi i/p}$.

Historically in the literature (cf. Borevich and Shafarevich: *Number Theory*, Academic Press, 1966, pp. 378–381), it was shown that if $\mathbb{Z}[\omega]$ is a UFD, then the theorem is true for p. But unfortunately, $\mathbb{Z}[\omega]$ is not always a UFD (see section 3). Actually, we have $\mathbb{Z}[\omega] = \mathbb{Q}(\omega) \cap \mathcal{A}$, where $\omega = e^{2\pi i/p}$, p an odd prime (see section 2). Though Fermat's last theorem was finally proved by Wiles in a very deep algebraic-geometric way in 1994, it was the ring $\mathcal{A}_K$ that started algebraic number theory and led to the ideal structure theory in both commutative and noncommutative algebra.

1. $\mathcal{A}_K$ is Normal and Free of Z-Rank $[K : \mathbb{Q}]$

Adopting the classical definition in number theory, if a field extension $\mathbb{Q} \subseteq K$, where $K \subset \mathbb{C}$, is finite dimensional, then K is called a *number field*. By Chapter 1 Theorem 3.11,

$$K = \mathbb{Q}(\alpha), \; \alpha \in K \text{ is algebraic over } \mathbb{Q}.$$

Using the notation as in the introduction of this chapter, let

$$\mathcal{A}_K = \mathcal{A} \cap K = \left\{ \beta \in K \; \middle| \; f(\beta) = 0 \text{ for a monic } f(x) \in \mathbb{Z}[x] \right\}.$$

Elements in $\mathcal{A}_K$ are called *algebraic integers*, and so, $\mathcal{A}_K$ is called the *ring of algebraic integers* of K.

Note that since the generator α of K is algebraic over $\mathbb{Q}$, there is some $0 \neq c \in \mathbb{Z}$ such that $c\alpha \in \mathcal{A}_K$ (later Exercise 1). It turns out that we may write

$$K = \mathbb{Q}(\vartheta), \; \vartheta \in \mathcal{A}_K.$$

Thus,

$$\mathbb{Z} \subseteq \mathbb{Z}[\vartheta] \subseteq \mathcal{A}_K \subset K = \mathbb{Q}(\vartheta).$$

1.1. Lemma With notation as above, $\mathcal{A}_K$ is equal to the integral closure of $\mathbb{Z}[\vartheta]$ in K, that is, $\mathcal{A}_K = \overline{\mathbb{Z}[\vartheta]}$. Hence $\mathcal{A}_K$ is a normal domain.

Proof This follows from Chapter 3 (section 1, exercise 5). □

1.2. Theorem (i) $\mathcal{A}_K$ is a finitely generated abelian group (or $\mathbb{Z}$-module). Therefore, $\mathcal{A}_K$ is normal and Noetherian.
(ii) $\mathcal{A}_K$ is a free abelian group of $\mathbb{Z}$-rank $[K : \mathbb{Q}]$.

Proof (i) Since $\mathbb{Z}$ is a normal Noetherian domain (a UFD is normal), $\mathbb{Z} \subset \mathbb{Q} \subseteq \mathbb{Q}(\vartheta)$, and $\mathcal{A}_K$ is the integral closure of $\mathbb{Z}$ in K, the assertion follows from Chapter 3 Theorem 3.2 and Chapter 1 Theorem 7.6.
(ii) By Chapter 1 (section 6, exercise 4), $\mathcal{A}_K$ is a free abelian group of finite rank. Note that any $\mathbb{Z}$-basis of $\mathcal{A}_K$ is also a $\mathbb{Q}$-basis for K (exercise 4). Hence the $\mathbb{Z}$-rank of $\mathcal{A}_K$ must be equal to $[K : \mathbb{Q}]$. □

In the literature, a $\mathbb{Z}$-basis of $\mathcal{A}_K$ is also called an *integral basis* of K, due to the fact that a $\mathbb{Z}$-basis of $\mathcal{A}_K$ is necessarily a $\mathbb{Q}$-basis for K (exercise 4).

Let $K = \mathbb{Q}(\vartheta)$ be as before, where $\vartheta \in \mathcal{A}_K$, $[K : \mathbb{Q}] = n$. From now on we let $p(x)$ be the minimal polynomial of ϑ over $\mathbb{Q}$.

1.3. Proposition For $\alpha \in K$, let $p_\alpha(x)$ be the minimal polynomial of α over $\mathbb{Q}$. Then $\alpha \in \mathcal{A}_K$ if and only if $p_\alpha(x) \in \mathbb{Z}[x]$. In particular, $p(x) \in \mathbb{Z}[x]$.

Proof The "if" part is clear, and the "only if" part follows from Chapter 1 Theorem 2.17 (note that $\alpha \in \mathcal{A}_K$ if $f(\alpha) = 0$ for some monic $f(x) \in \mathbb{Z}[x]$). □

Remark From Proposition 1.3 we derive immediately that $\mathbb{Q} \cap \mathcal{A} = \mathbb{Z}$. This recaptures the fact that $\mathbb{Z}$ is integrally closed in $\mathbb{Q}$.

Concerning integral bases of K, we first note an easy fact in view of Proposition 1.3.

Observation Let $\alpha \in \mathcal{A}_K$, then $p_\alpha(x) \in \mathbb{Z}[x]$. Since $p_\alpha(x)$ is monic, a division algorithm by $p_\alpha(x)$ in $\mathbb{Z}[x]$ plus the ring homomorphism $\mathbb{Z}[x] \to \mathbb{Z}[\alpha]$ with $x \mapsto \alpha$ yields $\mathbb{Z}[\alpha] = \sum_{i=0}^{n-1} \mathbb{Z}\alpha^i$, where $n = \deg p_\alpha(x)$. Consequently, $\mathbb{Z}[\alpha]$ is a free abelian group of rank n, and $\{1, \alpha, \alpha^2, ..., \alpha^{n-1}\}$ is a $\mathbb{Z}$-basis of $\mathbb{Z}[\alpha]$. So, if $\mathbb{Z}[\alpha] = \mathcal{A}_K$, then we have found a $\mathbb{Z}$-basis for $\mathcal{A}_K$ (in section 2 we will see that many number fields have this property).

However, finding an integral basis for an arbitrary number field K is by no means an easy job in algebraic number theory. As to this topic, a very useful invariant of K is introduced and discussed below.

Consider the n *distinct* zeros of $p(x)$ in $\mathbb{C}$ (Chapter 1 Theorem 3.6), say

$$\vartheta_1 = \vartheta, \vartheta_2, ..., \vartheta_n.$$

By Chapter 1 Proposition 5.1, there are exactly n *distinct* $\mathbb{Q}$-linear ring monomorphisms

$$\begin{aligned} \sigma_i : \ K = \mathbb{Q}(\vartheta) &\longrightarrow \mathbb{C} \\ \vartheta &\mapsto \vartheta_i \end{aligned}$$

$i = 1, ..., n$.

1.4. Definition Let $\sigma_1, ..., \sigma_n$ be as above. If $\{\alpha_1, ..., \alpha_n\}$ is a $\mathbb{Q}$-basis for

$K = \mathbb{Q}(\vartheta)$, then

$$\Delta[\alpha_1, ..., \alpha_n] = [\det(\sigma_i(\alpha_j))]^2$$

is called the *discriminant of the given basis*.

1.5. Lemma Let $\{\alpha_1, ..., \alpha_n\}$ and $\{\beta_1, ..., \beta_n\}$ be two $\mathbb{Q}$-bases of $K = \mathbb{Q}(\vartheta)$. Then

$$\Delta[\beta_1, ..., \beta_n] = [\det(c_{jk})]^2 \Delta[\alpha_1, ..., \alpha_n],$$

where $\beta_k = \sum_{j=1}^n c_{jk}\alpha_j$, $k = 1, .., n$, $c_{jk} \in \mathbb{Q}$.

Proof Exercise. □

1.6. Proposition With notation as above, the following hold:
(i) $0 \neq \Delta[\alpha_1, ..., \alpha_n] \in \mathbb{Q}$ for any $\mathbb{Q}$-basis $\{\alpha_1, ..., \alpha_n\}$ of $K = \mathbb{Q}(\vartheta)$.
(ii) If all $\sigma_i(\vartheta)$, $i = 1, ..., n$, are real numbers, then $\Delta[\alpha_1, ..., \alpha_n] > 0$ for any $\mathbb{Q}$-basis $\{\alpha_1, ..., \alpha_n\}$ of K.

Proof (i) Consider the $\mathbb{Q}$-basis $\{1, \vartheta, \vartheta^2, ..., \vartheta^{n-1}\}$ and note that $\sigma_i(\vartheta) = \vartheta_i$, $i = 1, ..., n$. Then

$$\Delta[1, \vartheta, \vartheta^2, ..., \vartheta^{n-1}] = \begin{vmatrix} 1 & \sigma_1(\vartheta) & \sigma_1(\vartheta^2) & \cdots & \sigma_1(\vartheta^{n-1}) \\ 1 & \sigma_2(\vartheta) & \sigma_2(\vartheta^2) & \cdots & \sigma_2(\vartheta^{n-1}) \\ \vdots & & & & \vdots \\ 1 & \sigma_n(\vartheta) & \sigma_n(\vartheta^2) & \cdots & \sigma_n(\vartheta^{n-1}) \end{vmatrix}^2$$

$$= \begin{vmatrix} 1 & \vartheta_1 & \vartheta_1^2 & \cdots & \vartheta_1^{n-1} \\ 1 & \vartheta_2 & \vartheta_2^2 & \cdots & \vartheta_2^{n-1} \\ \vdots & & & & \vdots \\ 1 & \vartheta_n & \vartheta_n^2 & \cdots & \vartheta_n^{n-1} \end{vmatrix}^2 = \left[\prod_{i<j} (\vartheta_i - \vartheta_j)\right]^2.$$

Since applying any permutation π of $\{\vartheta_1, ..., \vartheta_n\}$ to $\prod_{i<j}(\vartheta_i - \vartheta_j)$ is the same as applying π to $\det(\sigma_i(\vartheta^j))$, it interchanges rows of $\det(\sigma_i(\vartheta^j))$ and hence the sign of $\det(\sigma_i(\vartheta^j))$. But then $\det(\sigma_i(\vartheta^j))^2$ is symmetric on $\vartheta_1, ..., \vartheta_n$. It follows from Chapter 1 Corollary 7.2 that $\Delta[1, \vartheta, \vartheta^2, ..., \vartheta^{n-1}] \in \mathbb{Q}$. Since all ϑ_i's are distinct, $\Delta[1, \vartheta, \vartheta^2, ..., \vartheta^{n-1}] \neq 0$ and the conclusion (i) follows from Lemma 1.5.
(ii) This follows from part (i) and Lemma 1.5. □

1.7. Corollary If $\{\alpha_1, ..., \alpha_n\}$ is a $\mathbb{Q}$-basis of $K = \mathbb{Q}(\vartheta)$ in which all α_i's are algebraic integers, then $0 \neq \Delta[\alpha_1, ..., \alpha_n] \in \mathcal{A} \cap \mathbb{Q} = \mathbb{Z}$.

Proof This follows from Proposition 1.6(i) and the fact that all $\sigma_i(\alpha_j)$'s are algebraic integers. □

As argued in [ST], there is a direct proof of Theorem 1.2(ii) by using only discriminant.

1.8. Another proof of Theorem 1.2(ii) Since $\{1, \vartheta, \vartheta^2, ..., \vartheta^{n-1}\}$ is a $\mathbb{Q}$-basis of K consisting of algebraic integers and $0 \neq \Delta[1, \vartheta, ..., \vartheta^{n-1}] \in \mathbb{Z}$ by Corollary 1.7, we may select a $\mathbb{Q}$-basis $W = \{\xi_1, ..., \xi_n\}$ of algebraic integers for which $|\Delta[\xi_1, ..., \xi_n]|$ is the least. We claim that $\{\xi_1, ..., \xi_n\}$ is a $\mathbb{Z}$-basis for $\mathcal{A}_K$. To see this, note that since $\{\xi_1, ..., \xi_n\}$ is also $\mathbb{Z}$-linearly independent, we need only to show that, for $w \in \mathcal{A}_K$, if $w = a_1\xi_1 + \cdots + a_n\xi_n$ with $a_i \in \mathbb{Q}$ then $a_i \in \mathbb{Z}$.

If $a_1 \notin \mathbb{Z}$, then $a_1 = a + r$ where $a \in \mathbb{Z}$ and $0 < r < 1$. Define

$$\eta_1 = w - a\xi_1 = (a_1 - a)\xi_1 + a_2\xi_2 + \cdots + a_n\xi_n,$$
$$\eta_2 = \xi_2,$$
$$\vdots$$
$$\eta_n = \xi_n,$$

we have

$$\begin{pmatrix} \eta_1 \\ \eta_2 \\ \vdots \\ \eta_n \end{pmatrix} = \begin{pmatrix} a_1 - a & a_2 & \cdots & a_n \\ 0 & 1 & \cdots & 0 \\ \vdots & & \vdots & \\ 0 & 0 & \cdots & 1 \end{pmatrix} \begin{pmatrix} \xi_1 \\ \xi_2 \\ \vdots \\ \xi_n \end{pmatrix}.$$

Let P be the above $n \times n$ matrix. Then $\det(P) = a_1 - a = r > 0$. This shows that $\{\eta_1, ..., \eta_n\}$ is also a $\mathbb{Q}$-basis of K consisting of algebraic integers. Now

$$\Delta[\eta_1, ..., \eta_n] = (\det(P))^2 \Delta[\xi_1, ..., \xi_n]$$

$$= r^2 \Delta[\xi_1, ..., \xi_n] < \Delta[\xi_1, ..., \xi_n]$$

contradicts the choice of $\{\xi_1, ..., \xi_n\}$. Therefore $a_1 \in \mathbb{Z}$. Similarly, all $a_i \in \mathbb{Z}$, as desired. □

It is not difficult to verify that any two integral bases of K have the same discriminant (exercise 4). Thus, the discriminant of any integral basis of K is called the *discriminant of K*.

1.9. Proposition Let $\{\alpha_1, ..., \alpha_n\}$ be a $\mathbb{Q}$-basis of $K = \mathbb{Q}(\vartheta)$ with all $\alpha_i \in \mathcal{A}_K$. If $\Delta[\alpha_1, ..., \alpha_n]$ is square-free then $\{\alpha_1, ..., \alpha_n\}$ is an integral basis of K.

Proof Let $\{\beta_1, ..., \beta_n\}$ be an integral basis and

$$\alpha_i = \sum_{j=1}^{n} c_{ij}\beta_j, \ i = 1, ..., n, \ c_{ij} \in \mathbb{Z}.$$

By lemma 1.5, $\Delta[\alpha_1, ..., \alpha_n] = (\det(c_{ij}))^2 \Delta[\beta_1, ..., \beta_n]$. Since $\Delta[\alpha_1, ..., \alpha_n]$ is square-free, it follows that $\det(c_{ij}) = \pm 1$. This shows that the matrix (c_{ij}) is unimodular and hence $\{\alpha_1, ..., \alpha_n\}$ is a $\mathbb{Z}$-basis for $\mathcal{A}_K$ by Chapter 1 Lemma 6.5. $\square$

Example (i) Let $K = \mathbb{Q}(\sqrt{d})$, where $d \in \mathbb{Z}$ is square-free and $4|(d-1)$. Consider $\alpha = \frac{1}{2} + \frac{1}{2}\sqrt{d}$. Then, α has minimal polynomial $p_\alpha(x) = x^2 - x + \frac{1-d}{4}$ in $\mathbb{Z}[x]$ (Chapter 1 section 3, exercise 6). Hence $\alpha \in \mathcal{A}_K$, but $\alpha \notin \mathbb{Z}[\sqrt{d}]$ (why?). So $\mathcal{A}_K \neq \mathbb{Z}[\sqrt{d}]$. Note that $\{1, \alpha\}$ forms a $\mathbb{Q}$-basis of K and the $\mathbb{Q}$-linear ring monomorphisms $K \to \mathbb{C}$ are defined by $\sigma_1(\sqrt{d}) = \sqrt{d}$ and $\sigma_2(\sqrt{d}) = -\sqrt{d}$ respectively. We have

$$\Delta[1, \alpha] = \begin{vmatrix} 1 & \frac{1}{2} + \frac{1}{2}\sqrt{d} \\ 1 & \frac{1}{2} - \frac{1}{2}\sqrt{d} \end{vmatrix}^2 = d.$$

It follows from Proposition 1.9 that $\{1, \alpha\}$ is an integral basis of K, for d is square-free by the assumption.

In section 2, we will establish integral bases for quadratic number fields and cyclotomic fields.

We end this section by giving two easy but important facts concerning the trace function $T_{K/\mathbb{Q}}$ and the norm function $N_{K/\mathbb{Q}}$ on K (Chapter 1 section 5) determined by the n distinct $\mathbb{Q}$-linear ring monomorphisms σ_j: $K = \mathbb{Q}(\vartheta) \to \mathbb{C}$. With no confusion we write T and N in place of $T_{K/\mathbb{Q}}$

and $N_{K/\mathbb{Q}}$ respectively. By definition, for $\alpha \in K$,

$$\sum_{j=1}^{p-1} \sigma_j(\alpha) = T(\alpha) \in \mathbb{Q},$$

$$\prod_{j=1}^{p-1} \sigma_j(\alpha) = N(\alpha) \in \mathbb{Q}.$$

1.10. Lemma Let $K = \mathbb{Q}(\vartheta)$ and $\mathcal{A}_K$ be as before. With notation as above, the following hold for $\alpha \in \mathcal{A}_K$:
(i) $\sigma_j(\alpha) \in \mathcal{A} = \{\beta \in \mathbb{C} \mid \beta \text{ is an algebraic integer}\}$, $j = 1, ..., n$.
(ii) $T(\alpha), N(\alpha) \in \mathcal{A} \cap \mathbb{Q} = \mathbb{Z}$.

Proof Exercise.

Exercises
1. Show that if $\alpha \in \mathbb{C}$ is an algebraic element over $\mathbb{Q}$, then there is some $0 \neq c \in \mathbb{Z}$ such that $c\alpha$ is an algebraic integer.
2. Show that a $\mathbb{Z}$-basis of $\mathcal{A}_K$ is also a $\mathbb{Q}$-basis for the number field $K = \mathbb{Q}(\vartheta)$.
3. Complete the proof of Lemma 1.5.
4. Show that any two integral bases of the number field $K = \mathbb{Q}(\vartheta)$ have the same discriminant. (Hint: Use Lemma 1.5 and Chapter 1 Lemma 6.5.)
5. Show that if two number fields K_1 and K_2 are isomorphic, then they have the same discriminant.
6. Find an example to show that the converse of Proposition 1.9 is not true in general.
7. Let $K = \mathbb{Q}(\vartheta)$ be a number field where ϑ has the minimal polynomial $p(x)$ of degree n. Verify

$$\Delta[1, \vartheta, ..., \vartheta^{n-1}] = (-1)^{n(n-1)/2} \cdot N_{K/\mathbb{Q}}\left(\frac{dp(x)}{dx}(\vartheta)\right).$$

(Hint: See the proof of Proposition 1.6 and calculate the right-hand side of the equation.)
8. Complete the proof of Lemma 1.10 and answer whether $\sigma_j(\alpha) \in \mathcal{A}_K$ or not. (Hint: Consider $K = \mathbb{Q}(\vartheta)$ where ϑ is the real root of $x^3 - 2$.)

2. $\mathbb{Q}(\sqrt{d})$ and $\mathbb{Q}(\omega)$

This section aims to establish the integral bases for quadratic number fields and cyclotomic number fields.

(I) $\mathcal{A}_K$ in $K = \mathbb{Q}(\sqrt{d})$.
Let $K = \mathbb{Q}(\vartheta)$ be a number field with $\vartheta \in \mathcal{A}_K$ and $[K : \mathbb{Q}] = 2$. Then K is called a *quadratic number field*.

Since $\vartheta \in \mathcal{A}_K$, by Proposition 1.3 its minimal polynomial over $\mathbb{Q}$ is of the form
$$x^2 + bx + c, \quad b, c \in \mathbb{Z}.$$
Thus, $\vartheta = \frac{-b \pm \sqrt{b^2 - 4c}}{2}$. Factorizing $b^2 - 4c$ in $\mathbb{Z}$, we may write $b^2 - 4c = \ell^2 d$ where d is square-free. It follows that $\vartheta = \frac{-b \pm \ell \sqrt{d}}{2}$. Hence, we conclude that

- $K = \mathbb{Q}(\sqrt{d})$, where $d \in \mathbb{Z}$ is square-free.

2.1. Theorem Let $d \in \mathbb{Z}$ be square-free, $K = \mathbb{Q}(\sqrt{d})$. The following hold:
(i) If $4 \nmid (d-1)$ then $\mathcal{A}_K = \mathbb{Z}[\sqrt{d}]$. Hence $\{1, \sqrt{d}\}$ is an integral basis for K.
(ii) If $4 | (d-1)$, then $\mathcal{A}_K = \mathbb{Z}\left[\frac{1}{2} + \frac{1}{2}\sqrt{d}\right]$. Hence $\{1, \frac{1}{2} + \frac{1}{2}\sqrt{d}\}$ is an integral basis for K.

Proof Let $\alpha \in \mathcal{A}_K$. Then $\alpha = r + s\sqrt{d}$, $r, s \in \mathbb{Q}$. Write
$$\alpha = \frac{a + b\sqrt{d}}{c} \quad \text{with } a, b, c \in \mathbb{Z}, \ c > 0.$$

(a) If $b = 0$, then $\alpha \in \mathcal{A}_K$ implies $\alpha \in \mathbb{Z} \subset \mathbb{Z}[\sqrt{d}]$.
(b) If $a = 0$, $b \neq 0$, since d is square-free, it follows from Proposition 1.3 that $\alpha \in \mathcal{A}_K$ implies $\alpha \in \mathbb{Z}[\sqrt{d}]$.
(c) If $a \neq 0$, $b \neq 0$, we may assume $\gcd(a, b, c) = 1$. The final conclusion will follow from a careful analysis of the coefficients of the minimal polynomial of α:
$$p_\alpha(x) = x^2 - \frac{2a}{c}x + \frac{a^2 - b^2 d}{c^2}.$$
By Proposition 1.3,
$$\alpha \in \mathcal{A}_K \text{ if and only if } \frac{2a}{c}, \frac{a^2 - b^2 d}{c^2} \in \mathbb{Z}.$$

But $\frac{a^2-b^2d}{c^2} \in \mathbb{Z}$ implies $\gcd(a,c) = 1$ because d is square-free and $\gcd(a,b,c) = 1$. Thus

$$\frac{2a}{c} \in \mathbb{Z} \text{ implies } c = 1 \text{ or } c = 2.$$

If $c = 1$, then $a + b\sqrt{d} = \alpha \in \mathbb{Z}[\sqrt{d}]$.
If $c = 2$, then a and b are odd as $\gcd(a,c) = 1$, $\frac{a^2-b^2d}{c^2} \in \mathbb{Z}$, and $\gcd(a,b,c) = 1$. This implies $\alpha \in \mathbb{Z}\left[\frac{1}{2} + \frac{1}{2}\sqrt{d}\right]$, $4|(a^2-1)$, and $4|(b^2-1)$. Writing

$$\frac{a^2 - b^2 d}{c^2} = \frac{(a^2-1) - (b^2-1)d - (d-1)}{4},$$

then $\frac{a^2-b^2d}{c^2} \in \mathbb{Z}$ implies $4|(d-1)$.

In conclusion,
if $4 \nmid (d-1)$, then $c = 1$ and $\mathcal{A}_K = \mathbb{Z}[\sqrt{d}]$;
if $4|(d-1)$, then $\beta = \frac{1}{2} + \frac{1}{2}\sqrt{d}$ has minimal polynomial $x^2 - x + \frac{1-d}{4} = p_\beta(x) \in \mathbb{Z}[x]$ and hence $\beta \in \mathcal{A}_K$. But clearly $\beta \notin \mathbb{Z}[\sqrt{d}]$. Therefore, $c = 2$ and $\mathcal{A}_K = \mathbb{Z}\left[\frac{1}{2} + \frac{1}{2}\sqrt{d}\right]$. □

(II) $\mathcal{A}_K$ in $K = \mathbb{Q}(\omega)$.
For $n \geq 1$, consider

$$U_n = \left\{\omega \in \mathbb{C} \;\middle|\; \omega^n - 1 = 0\right\}.$$

By Chapter 1 Proposition 3.4, U_n is a cyclic subgroup of $\mathbb{C}^\times$. U_n is called the *group of the nth roots of unity*. If $\omega \in U_n$ is a generator of U_n, i.e., $U_n = \langle \omega \rangle$, then ω is called a *primitive nth root of unity*. For instance, $\omega = e^{2\pi i/n} = \cos\frac{2\pi}{n} + i\sin\frac{2\pi}{n}$. From a first course of group theory one knows that there are $\varphi(n) = n\prod_{i=1}^{r}\left(1 - \frac{1}{p_i}\right)$ primitive roots of unity, where $n = p_1^{n_1} \cdots p_r^{n_r}$ is the factorization of n in $\mathbb{Z}$ and $\varphi(n)$ is the Euler's φ-function.

If $\omega \in U_n$, the number field $K = \mathbb{Q}(\omega)$ is called a *cyclotomic field*. That is, the roots of unity are the vertices of a regular n-gon located in the unit circle $\{\alpha \in \mathbb{C} \mid |\alpha| = 1\}$ of the complex α-plane with one vertex of the n-gon at $\alpha = 1$.

In view of the example given in the introductory part of this chapter, we restrict our attention to $n = p$, a *prime number*. If $p = 2$, then $U_2 = \{1, -1\}$ and $\mathbb{Q}(\omega) = \mathbb{Q}$. So we assume from now on that

- p is an odd prime number.

2.2. Lemma Let $\omega \in U_p$, $\omega \neq 1$. Then the minimal polynomial $p(x)$ of ω is

$$x^{p-1} + x^{p-2} + \cdots + x + 1.$$

Proof Write $f(x) = x^{p-1} + x^{p-2} + \cdots + x + 1$. Note that $\omega^p - 1 = 0$. We have $p(x)|(x^p - 1)$. Since $x^p - 1 = (x-1)f(x)$ and $\omega \neq 1$, it follows that $p(x)|f(x)$. But $f(x)$ is irreducible by Chapter 1 Corollary 2.22. Hence $p(x) = f(x)$. □

Let $\omega \in U_n$ be a primitive pth root of unity. Then $U_p = \langle \omega \rangle = \{1, \omega, \omega^2, ..., \omega^{p-1}\}$, and by Lemma 2.1, in $\mathbb{C}$

(1) $$p(x) = (x - \omega)\left(x - \omega^2\right) \cdots \left(x - \omega^{p-1}\right).$$

Moreover, for $K = \mathbb{Q}(\omega)$, $[K : \mathbb{Q}] = p - 1$, $\{1, \omega, ..., \omega^{p-2}\}$ is a $\mathbb{Q}$-basis of K, and all $\mathbb{Q}$-linear ring monomorphisms $K \to \mathbb{C}$ are given by

(2) $$\sigma_j : \begin{array}{c} K \longrightarrow \mathbb{C} \\ \omega \mapsto \omega^j \end{array} \quad j = 1, 2, ..., p-1.$$

Our aim is to show $\mathcal{A}_K = \mathbb{Z}[\omega]$ by making use of the trace function $T_{K/\mathbb{Q}}$ and the norm function $N_{K/\mathbb{Q}}$ on K (Chapter 1 section 5) determined by σ_j described in (2) above. As in Lemma 1.10, we write T and N in place of $T_{K/\mathbb{Q}}$ and $N_{K/\mathbb{Q}}$ respectively.

2.3. Lemma With notation as above, the following hold:
(i) For $\alpha \in \mathcal{A}_K$, $T(\alpha), N(\alpha) \in \mathbb{Z}$.
(ii) For $i = 1, ..., p-1$, $T(\omega^i) = -1$, $N(\omega^i) = 1$.
(iii) For $a \in \mathbb{Q}$, $T(a) = (p-1)a$, $N(a) = a^{p-1}$.
(iv) For $0 \neq s \in \mathbb{Z}$, $s = ph + r$, $0 \leq r \leq p-1$,

$$T(\omega^s) = \begin{cases} p-1, & \text{if } r = 0, \\ -1, & \text{if } r \neq 0. \end{cases}$$

$$N(\omega^s) = 1.$$

(v) For $a_0 + a_1\omega + \cdots + a_{p-2}\omega^{p-2} = \alpha \in K$,

$$T(\alpha) = (p-1)a_0 - \sum_{i=1}^{p-2} a_i = pa_0 - \sum_{i=0}^{p-2} a_i.$$

(vi) $N(1-\omega) = p = T(1-\omega) = T(1-\omega^2) = \cdots = T(1-\omega^{p-1})$.

Proof (i) By Lemma 1.10, if $\alpha \in \mathcal{A}_K$ then $T(\alpha), N(\alpha) \in \mathcal{A} \cap \mathbb{Q} = \mathbb{Z}$.
(ii) By the foregoing (1)–(2), $T(\omega^i) = p(\omega) - 1 = -1$, and $N(\omega^i) = p(0) = 1$.
(iii) Since all σ_j's are $\mathbb{Q}$-linear, this follows from the definition of T and N.
(iv) This follows from (ii) and (iii).
(v) Since T is a linear function by Chapter 1 Proposition 5.4, this follows from (ii)–(iii).
(vi) Note that by previous (1)–(2), $N(1-\omega) = \prod_{j=1}^{p-1}(1-\omega^j) = p(1) = p$.
That $T(1-\omega) = T(1-\omega^2) = \cdots = T(1-\omega^{p-1}) = p$ follows from (iii)–(iv).
□

2.4. Proposition For $K = \mathbb{Q}(\omega)$, where ω is a primitive pth root of unity, the following hold:
(i) Let $\langle 1-\omega \rangle$ be the ideal of $\mathcal{A}$ generated by $1-\omega$, where $\mathcal{A} = \{\beta \in \mathbb{C} \mid \beta$ is an algebraic integer$\}$. Then $\langle 1-\omega \rangle \cap \mathbb{Z} = p\mathbb{Z}$.
(ii) For $\alpha \in \mathcal{A}_K$, $T((1-\omega)\alpha) \in p\mathbb{Z}$.

Proof (i) By Lemma 2.3(vi),

$$(*) \quad p = N(1-\omega) = (1-\omega)(1-\omega^2)\cdots(1-\omega^{p-1}) \in \langle 1-\omega \rangle \cap \mathbb{Z},$$

and hence $p\mathbb{Z} \subseteq \langle 1-\omega \rangle \cap \mathbb{Z}$. Note that $p\mathbb{Z}$ is a maximal ideal of $\mathbb{Z}$. If $\langle 1-\omega \rangle \cap \mathbb{Z} \neq p\mathbb{Z}$, then $\langle 1-\omega \rangle \cap \mathbb{Z} = \mathbb{Z}$. This implies that $1-\omega$ is a unit of $\mathcal{A}$. Consequently, all $1-\omega^j$, $j = 1,...,p-1$, and also p, are units of $\mathcal{A}$ by the formula $(*)$ above. Thus, $p^{-1} \in \mathcal{A} \cap \mathbb{Q} = \mathbb{Z}$, that is impossible. This shows that $\langle 1-\omega \rangle \cap \mathbb{Z} = p\mathbb{Z}$.
(ii) Let $\langle 1-\omega \rangle$ be as in part (i) and let $a_0 + a_1\omega + \cdots + a_{p-2}\omega^{p-2} = \alpha \in \mathcal{A}_K$. Then by Lemma 1.10,

$$\sigma_j((1-\omega)\alpha) = (1-\omega^j)\sigma_j(\alpha) \in \langle 1-\omega \rangle$$

for $j = 1,...,p-1$. Thus, $\sum_{j=1}^{p-1} \sigma_j((1-\omega)\alpha) = T((1-\omega)\alpha) \in \langle 1-\omega \rangle \cap \mathbb{Z} = p\mathbb{Z}$ by (i) and Lemma 2.3(i).
□

Remark In Proposition 2.4 the reason for using the ring $\mathcal{A}$ instead of $\mathcal{A}_K$ is that we do not know if $\sigma_j(\alpha) \in \mathcal{A}_K$ (section 1, exercise 8).

2.5. Theorem For $K = \mathbb{Q}(\omega)$, where ω is a primitive pth root of unity, we have $\mathcal{A}_K = \mathbb{Z}[\omega]$. Hence $\{1, \omega, ..., \omega^{p-2}\}$ is an integral basis of K.

Proof Let $a_0 + a_1\omega + \cdots + a_{p-2}\omega^{p-2} = \alpha \in \mathcal{A}_K$ where $a_i \in \mathbb{Q}$. Then

$$(1-\omega)\alpha = a_0(1-\omega) + a_1(\omega - \omega^2) + \cdots + a_{p-2}(\omega^{p-2} - \omega^{p-1}),$$

and by Lemma 2.3 and Proposition 2.4(ii),

$$T((1-\omega)\alpha) = a_0 T(1-\omega) = a_0 p \in p\mathbb{Z}.$$

Hence $a_0 \in \mathbb{Z}$. Note that $\omega^{-1} = \omega^{p-1} \in \mathcal{A}_K$. It follows that

$$(\alpha - a_0)\omega^{-1} = a_1 + a_2\omega + \cdots + a_{p-2}\omega^{p-3}.$$

Proceeding as with a_0, it turns out that $a_1 \in \mathbb{Z}$. After repeating a similar argumentation $p-1$ times successively, we conclude that each $a_i \in \mathbb{Z}$. This shows that $\alpha \in \mathbb{Z}[\omega]$.

Exercises
1. Let p be an odd prime number and $K = \mathbb{Q}(\omega)$, where ω is a primitive pth root of unity. Use the integral basis $\{1, \omega, ..., \omega^{p-2}\}$ of K to show that K has the discriminant

$$\Delta = (-1)^{(p-1)(p-2)/2} \cdot N\left(\frac{dp(x)}{dx}(\omega)\right) = (-1)^{(p-1)/2} \cdot p^{p-2}.$$

(Hint: Use section 1, exercise 7.)

2. Let p be a prime number, $r \geq 1$, and ω a primitive p^rth root of unity. Show that the minimal polynomial of ω over $\mathbb{Q}$ is

$$x^{p^{r-1}(p-1)} + x^{p^{r-1}(p-2)} + \cdots + x^{p^{r-1}} + 1 = \frac{x^{p^r} - 1}{x^{p^{r-1}} - 1}.$$

Furthermore, use a similar argumentation of this section to show that for $K = \mathbb{Q}(\omega)$ the equality $\mathcal{A}_K = \mathbb{Z}[\omega]$ also holds. Can this be generalized to an arbitrary positive integer $n = p_1^{r_1} p_2^{r_2} \cdots p_s^{r_s}$ (a product of primes)?

3. Factorization of Elements in $\mathcal{A}_K$

This section specifies examples of $\mathcal{A}_K$ in which some are UFDs and some are not UFDs.

Let $K = \mathbb{Q}(\vartheta)$ be a number field, where $\vartheta \in \mathcal{A}_K$, $[K : \mathbb{Q}] = n$. Since $\mathcal{A}_K$ is a Noetherian domain (Theorem 1.2), factorization into irreducible elements is feasible in $\mathcal{A}_K$ (Chapter 1 Proposition 2.5). In order to realize the factorization of elements in $\mathcal{A}_K$, it is necessary to know how to recognize units and irreducible elements in $\mathcal{A}_K$.

As before we write $U(\mathcal{A}_K)$ for the multiplicative group of units in $\mathcal{A}_K$; and as before, for $\alpha \in \mathcal{A}_K$, we write $T(\alpha)$ for the trace of α and $N(\alpha)$ for the norm of α. By Lemma 1.10, $T(\alpha)$, $N(\alpha) \in \mathbb{Z}$.

3.1. Lemma For $\alpha, \beta \in \mathcal{A}_K$, the following hold:
(i) $\alpha \in U(\mathcal{A}_K)$ if and only if $N(\alpha) = \pm 1$.
(ii) If α and β are associates to each other (see Chapter 1 Definition 2.1), then $N(\alpha) = \pm N(\beta)$.
(iii) If $|N(\alpha)|$ is a prime number in $\mathbb{Z}$, then α is an irreducible element in $\mathcal{A}_K$. (The converse of this statement is not true as shown in later Example (ii).)

Proof (i) If $\alpha, \alpha^{-1} \in \mathcal{A}_K$, then $\alpha\alpha^{-1} = 1$ implies $N(\alpha\alpha^{-1}) = N(\alpha)N(\alpha^{-1}) = 1$, and hence $N(\alpha) = \pm 1$. Conversely, if $N(\alpha) = \sigma_1(\alpha)\sigma_2(\alpha)\cdots\sigma_n(\alpha) = \pm 1$, where $\sigma_1, ..., \sigma_n$ are all n distinct $\mathbb{Q}$-linear ring monomorphisms $K \to \mathbb{C}$, then, assuming $\sigma_1(\alpha) = \alpha$ without loss of generality, we have $\pm 1 = \alpha\sigma_2(\alpha)\cdots\sigma_n(\alpha)$. By Proposition 1.10(i), $\alpha^{-1} = \pm\sigma_2(\alpha)\cdots\sigma_n(\alpha) \in \mathcal{A} \cap K = \mathcal{A}_K$, that is, $\alpha \in U(\mathcal{A}_K)$.
(ii) This follows from (i).
(iii) Suppose $N(\alpha) = p$ is a prime number in $\mathbb{Z}$. If $\alpha = \beta\xi$ in $\mathcal{A}_K$, then $p = |N(\alpha)| = |N(\beta)||N(\xi)|$. Thus, $p|N(\beta)$ or $p|N(\xi)$. If $p|N(\beta)$, then $N(\xi) = \pm 1$ and $\xi \in U(\mathcal{A}_K)$ by (i). Similarly, if $p|N(\xi)$ then $\beta \in U(\mathcal{A}_K)$. This shows that α has no proper divisors in $\mathcal{A}_K$. □

The proposition given below finds $U(\mathcal{A}_K)$ in a quadratic field K of complex numbers.

3.2. Proposition Let $d \in \mathbb{Z}$, $d < 0$ and square-free, $K = \mathbb{Q}(\sqrt{d})$. Write $i = \sqrt{-1}$.
(i) $U(\mathcal{A}_K) = \{\pm 1, \pm i\}$ if $d = -1$.
(ii) $U(\mathcal{A}_K) = \{\pm 1, \pm \omega, \pm \omega^2\}$ if $d = -3$, where $\omega = e^{2\pi i/3}$.
(iii) $U(\mathcal{A}_K) = \{\pm 1\}$ if $d < -3$.

Proof If $\alpha \in U(\mathcal{A}_K)$ then $N(\alpha) = \pm 1$ by Lemma 3.1(i). Write $\alpha = r + s\sqrt{d}$

in K. Since $d < 0$, it follows from Chapter 1 (section 5, exercise 3) that

(0) $\qquad \pm 1 = N(\alpha) = r^2 - s^2 d$ implies $N(\alpha) = +1$.

If $4 \nmid (d-1)$, then $\alpha = a + b\sqrt{d}$ with $a, b \in \mathbb{Z}$ by Theorem 2.1(i), and by (0) above,

(1) $\qquad N(\alpha) = a^2 - b^2 d = 1.$

If $4 | (d-1)$, then

$$\alpha = a + b\left(\frac{1}{2} + \frac{1}{2}\sqrt{d}\right) = \frac{2a+b}{2} + \frac{b}{2}\sqrt{d}, \ a, b \in \mathbb{Z},$$

and by (0) above,

(2) $\qquad N(\alpha) = 1$ implies $(2a+b)^2 - b^2 d = 4.$

If $d = -1$, then (i) follows from (1) above.
If $d = -3$, then by (2) above, $(2a+b)^2 + 3b^2 = 4$ and this yields

$$b = 0 \text{ implies } a = \pm 1;$$

or

$$b = \pm 1 \text{ implies } (2a+b)^2 = 1 \text{ implies } \begin{cases} 2a+1 = +1 \text{ implies } a = 0 \\ 2a+1 = -1 \text{ implies } a = -1 \\ 2a-1 = +1 \text{ implies } a = 1 \\ 2a-1 = -1 \text{ implies } a = 0. \end{cases}$$

This proves (ii).

If $d < -3$ and $4 \nmid (d-1)$, then by (1) above, $b = 0$ and $a = \pm 1$. If $d < -3$ and $4 | (d-1)$, then by (2) above, $b = 0$ and $a = \pm 1$. This proves (iii). $\qquad \square$

Remark If $d \geq 2$ and d is square-free, then $K = \mathbb{Q}(\sqrt{d})$ is a subfield of $\mathbb{R}$ and $U(\mathcal{A}_K)$ is no longer necessarily finite. For instance, in $K = \mathbb{Q}(\sqrt{2})$, $1 + \sqrt{2} = \alpha \in U(\mathcal{A}_K)$ because $N(\alpha) = -1$ (indeed, $(1+\sqrt{2})(-1+\sqrt{2}) = 1$); and since $\alpha^n > 1$ for $n \geq 1$, it follows that $\{\pm \alpha^n\}_{n \geq 1} \subset U(\mathcal{A}_K)$, i.e., $U(\mathcal{A}_K)$ is infinite. For square-free $d \geq 2$, $K = \mathbb{Q}(\sqrt{d})$, a general fact is that the *positive units* of $\mathcal{A}_K$ form a multiplicative group isomorphic to $\mathbb{Z}$ (cf. [Sam] section 4.6, Proposition 1).

For a general structure theory on $U(\mathcal{A}_K)$, where $K = \mathbb{Q}(\vartheta)$ is an arbitrary number field, we refer the reader to the famous Dirichlet units theorem (cf. [Sam]).

The following proposition enables us to see how the norm function on a number field K can help to construct a Euclidean function on $\mathcal{A}_K$.

3.3. Theorem Let $K = \mathbb{Q}(\vartheta)$ be a number field, $\mathcal{A}_K^\times = \mathcal{A}_K - \{0\}$. For $\alpha \in K$, let $N(\alpha)$ be the norm of α. Then the correspondence

$$\phi: \mathcal{A}_K^\times \longrightarrow \mathbb{N}$$
$$\alpha \mapsto |N(\alpha)|$$

defines a Euclidean function on $\mathcal{A}_K$ if and only if for any $\eta \in K$ there exists $\xi \in \mathcal{A}_K$ such that $|N(\eta - \xi)| < 1$.

Proof If the function has the stated property, we show that ϕ satisfies the conditions (i)–(ii) of Chapter 1 Definition 1.11. By Lemma 1.10, $N(\alpha) \in \mathbb{Z}$ for $\alpha \in \mathcal{A}_K$. If $\alpha | \alpha'$ in $\mathcal{A}_K$ then $\alpha' = \alpha\mu$ for some $\mu \in \mathcal{A}_K$. Thus, $|N(\alpha')| = \phi(\alpha') = \phi(\alpha)\phi(\mu) = |N(\alpha)| \cdot |N(\mu)|$ and hence $\phi(\alpha) \leq \phi(\alpha')$. This shows that Chapter 1 Definition 2.11(i) holds. To see that Chapter 1 Definition 2.11(ii) holds as well, for $\alpha, \beta \in \mathcal{A}_K^\times$, set $\eta = \frac{\alpha}{\beta}$. Then there is $\xi \in \mathcal{A}_K$ such that $|N(\eta - \xi)| < 1$, that is,

$$\left| N\left(\frac{\alpha - \beta\xi}{\beta}\right) N(\beta) \right| < |N(\beta)| \text{ or } |N(\alpha - \beta\xi)| < |N(\beta)|.$$

Put $r = \alpha - \beta\xi$. Then $\alpha = \beta\xi + r$ where $\phi(r) < \phi(\beta)$, as desired.

Conversely, suppose that ϕ defines a Euclidean function. Let $\eta \in K$. Then there is some $c \in \mathbb{Z}$ such that $c\eta \in \mathcal{A}_K$ (section 1, Exercise 1). Now for $\alpha = c\eta$, $\beta = c$, we have $\alpha = \beta\mu + r$, where $\mu, r \in \mathcal{A}_K$ with the property that $r = 0$ or $\phi(r) < \phi(\beta)$. If $r = 0$, then $\alpha = \beta\mu$, i.e., $c\eta = c\mu$ and $\eta = \mu \in \mathcal{A}_K$. In this case, taking $\xi = \mu$ fulfils the job. If $r \neq 0$, then $c \neq 0$ implies

$$\phi(r) = |N(r)| = |N(\alpha - \beta\mu)| = |N(c\eta - c\mu)| < |N(c)| = \phi(\beta).$$

After multiplying both sides of the formula above by $|N(\frac{1}{c})|$, we have $|N(\eta - \mu)| < 1$. This shows that the choice $\xi = \mu$ is a desired object. □

Example Let $r \in \mathbb{Q}$ be a rational number and $[r]$ the largest integer smaller than or equal to r. Then $[r] \leq r < [r] + 1$. Thus,

$$\text{either } |r - [r]| \leq \frac{1}{2}, \text{ or } |r - [r] - 1| \leq \frac{1}{2}.$$

With the help of this observation, the following two examples are easily presented.

(i) Let $K = \mathbb{Q}(\sqrt{d})$ with $d = 2, 3$. Then $\mathcal{A}_K = \mathbb{Z}[\sqrt{d}]$. For any $r + s\sqrt{d} = \eta \in K$ and $a + b\sqrt{d} = \xi \in \mathcal{A}_K$, $|N(\eta - \xi)| = |(r-a)^2 - d(s-b)^2|$. So it is possible to choose $a + b\sqrt{d} = \xi \in \mathcal{A}_K$ such that $|N(\eta - \xi)| = |(r-a)^2 - d(s-b)^2| < 1$. This shows that $\mathbb{Z}[\sqrt{d}]$ is Euclidean and hence a UFD.

(ii) Let $K = \mathbb{Q}(i)$ where $i = \sqrt{-1}$. Then $\mathcal{A}_K = \mathbb{Z}[i]$. For any $r + si = \eta \in K$, we may find $a, b \in \mathbb{Z}$, such that $(r-a)^2, (s-b)^2 \leq \frac{1}{4}$. Thus, for $\xi = a + bi$ we have $|N(\eta - \xi)| = |(r-a)^2 + (s-b)^2| < 1$. This shows that $\mathbb{Z}[i]$ is Euclidean and hence a UFD.

Example (iii) Let $K = \mathbb{Q}(\sqrt{-5})$. Then $\mathcal{A}_K = \mathbb{Z}[\sqrt{-5}]$, and by Proposition 3.2, $U(\mathcal{A}_K) = \{\pm 1\}$. Using Proposition 3.1, one checks that $2, 3, 1 \pm \sqrt{-5}$ are irreducible elements but not primes in $\mathcal{A}_K$, because

$$6 = 2 \cdot 3 = \left(1 + \sqrt{-5}\right) \cdot \left(1 - \sqrt{-5}\right).$$

This shows that $\mathcal{A}_K$ is not a UFD.

Moreover, $N(1 + \sqrt{-5}) = 6$. This shows that the converse of Proposition 3.1(iii) is not true.

Without proof we mention the following result from algebraic number theory (for instance, cf. [ST]).

3.4. Theorem Let $d \in \mathbb{Z}$ be square-free, $K = \mathbb{Q}(\sqrt{d})$. If $d < 0$, then $\mathcal{A}_K$ is Euclidean if and only if $d = -1, -2, -3, -7, -11$. And for each of these listed numbers, the Euclidean function is the one defined in Theorem 3.3.

□

Finally, the reader is referred to [Edw] for a detailed argumentation about the fact that for the cyclotomic field $K = \mathbb{Q}(\omega)$, where $\omega = e^{2\pi i/23}$, $\mathcal{A}_K$ is not a UFD.

Exercises

1. Let $K = \mathbb{Q}(i)$, $i = \sqrt{-1}$. Which of 2, 3, 5 and 7 are reducible in $\mathcal{A}_K$? Which of $1+i$, $3-7i$, and $-4+5i$ are irreducible in $\mathcal{A}_K$.
2. Let $K = \mathbb{Q}(\sqrt{3})$.
   (a) Show that $\alpha = 5 + 3\sqrt{3}$ is irreducible in $\mathcal{A}_K$, and that $\beta = 7 + 4\sqrt{3}$ is a unit in $\mathcal{A}_K$.
   (b) Use part (a) to show that $71 + 41\sqrt{3}$ and $5 + 3\sqrt{3}$ are associates to each other.
   (c) Is 2 a prime in $\mathcal{A}_K$? (Hint: Consider $6 = 2 \cdot 3 = (3 + \sqrt{3})(3 - \sqrt{3})$, $2 = (5 + 3\sqrt{3})(-5 + 3\sqrt{3})$.)
3. Use the factorizations

$$18 = 2 \cdot 3 \cdot 3 = (1 + \sqrt{-17}) \cdot (1 - \sqrt{-17}),$$

$$6 = 2 \cdot 3 = (4 + \sqrt{10}) \cdot (4 - \sqrt{10}), \text{ and}$$

$$10 = 2 \cdot 5 = (5 + \sqrt{15}) \cdot (5 - \sqrt{15})$$

to show that $\mathcal{A}_K$ is not a UFD for $K = \mathbb{Q}(\sqrt{-17})$, $K = \mathbb{Q}(\sqrt{10})$ and $K = \mathbb{Q}(\sqrt{15})$, respectively.

4. Show that $\mathcal{A}_K$ is Euclidean where $K = \mathbb{Q}(\sqrt{5})$. (Hint: Consider the function $\phi \colon \mathcal{A}_K^\times \to \mathbb{N}$ with

$$\phi\left(a + b\left(\frac{1}{2} + \frac{\sqrt{5}}{2}\right)\right) = |a^2 + ab - b^2|,$$

and show that for any $r + s\sqrt{5} = \eta \in K$, there is $a + b(\frac{1}{2} + \frac{1}{2}\sqrt{5}) = \xi \in \mathcal{A}_K$ such that $\phi(\eta - \xi) \leq \frac{3}{4}$.)

4. From $\mathcal{A}_K$ to Dedekind Domains

Let $K = \mathbb{Q}(\vartheta)$ be a number field, $[K : \mathbb{Q}] = n$, and $\mathcal{A}_K$ its ring of algebraic integers. We have seen in section 3 that $\mathcal{A}_K$ is not necessarily a UFD. In this section we clarify how far is $\mathcal{A}_K$ from being a UFD, by studying the structure of ideals in $\mathcal{A}_K$. More precisely, we show that every nonzero ideal I of $\mathcal{A}_K$ can be expressed as a product of finitely many prime (indeed maximal) ideals uniquely up to the order of factors, that I is generated by at most two elements, and that $\mathcal{A}_K$ is a UFD if and only if it is a PID.

We start with a useful fact.

4.1. Lemma Let I be an ideal of $\mathcal{A}_K$. If $0 \neq \alpha \in I$, then $0 \neq N(\alpha) \in I \cap \mathbb{Z}$, where $N(\alpha)$ stands for the norm of α.

Proof Let $\sigma_1, ..., \sigma_n$ be the n distinct $\mathbb{Q}$-linear ring monomorphisms $K \to \mathbb{C}$, where $\sigma_1(\alpha) = \alpha$. By Lemma 1.10, $\sigma_1(\alpha)\sigma_2(\alpha) \cdots \sigma_n(\alpha) = N(\alpha) \in \mathbb{Z}$, and since $\alpha \neq 0$, $\sigma_2(\alpha) \cdots \sigma_n(\alpha) = \alpha^{-1} N(\alpha) \in K \cap \mathcal{A} = \mathcal{A}_K$. It follows that $N(\alpha) \in I \cap \mathbb{Z}$. □

4.2. Theorem (i) Every nonzero prime ideal of $\mathcal{A}_K$ is a maximal ideal (hence nonzero and minimal).
(ii) If P is a nonzero prime ideal of $\mathcal{A}_K$, then the localization $(\mathcal{A}_K)_P$ of $\mathcal{A}_K$ at P is a DVR.

Proof (i) Let P be a nonzero prime ideal of $\mathcal{A}_K$ and $0 \neq \alpha \in P$. By Lemma 4.1, $0 \neq N(\alpha) \in P \cap \mathbb{Z}$. Let $N(\alpha) = q$ and $\langle q \rangle$ the ideal of $\mathcal{A}_K$ generated by q. Then $\langle q \rangle \subseteq P$, and we have the natural onto ring homomorphism $\mathcal{A}_K / \langle q \rangle \to \mathcal{A}_K / P$. If $\{\xi_1, ..., \xi_n\}$ is an integral basis of K (or by Theorem 1.2, a $\mathbb{Z}$-basis of $\mathcal{A}_K$), then it is easy to see that every element of $\mathcal{A}_K / \langle q \rangle$ is of the form

$$(a_1 \xi_1 + a_2 \xi_2 + \cdots + a_n \xi_n) + \langle q \rangle, \ 0 \leq a_i \leq q-1, \ i = 1, ..., n,$$

that is, $\mathcal{A}_K / \langle q \rangle$ has at most q^n elements. Consequently, $\mathcal{A}_K / P$ is a finite domain and hence a field (Chapter 1 Proposition 0.1). It follows from Chapter 2 Proposition 1.2(ii) that P is maximal.
(ii) This follows from part (i) and Chapter 3 Corollary 4.6. □

4.3. Proposition Let I be a nonzero ideal of $\mathcal{A}_K$. Then there exist prime ideals $P_1, ..., P_s$ such that $P_1 \cdots P_s \subseteq I$.

Proof If I is a prime ideal, it is done. Suppose that I is not a prime ideal. If the assertion was not true for I, then the set

$$\Omega = \{\text{ideals of } \mathcal{A}_K \text{ not containing any product of prime ideals}\} \neq \emptyset.$$

Since $\mathcal{A}_K$ is Noetherian, Ω has a maximal member with respect to $\subseteq$, say M. Thus M is not a prime ideal, and there are ideals J_1, J_2 of $\mathcal{A}_K$ with $J_1 J_2 \subseteq M$ but $J_1 \not\subseteq M$, $J_2 \not\subseteq M$. Put $W_1 = J_1 + M$, $W_2 = J_2 + M$. Then $M \subset W_1$, $M \subset W_2$, and by the choice of M, $W_1, W_2 \notin \Omega$. Thus, both W_1 and W_2 contain some product of prime ideals. But $W_1 W_2 \subseteq M$ and this is a contradiction. Therefore $\Omega = \emptyset$ and the assertion is proved. □

Our aim is to show that every nonzero ideal I is a product of finitely many prime ideals. In doing so, we need to study special $\mathcal{A}_K$-submodules of K. Note that since $\mathcal{A}_K$ is a subring of K, K forms an $\mathcal{A}_K$-module in a natural way (Chapter 1 section 7).

4.4. Definition An $\mathcal{A}_K$-submodule J of K is called a *fractional ideal* of $\mathcal{A}_K$ if there exists some $0 \neq c \in \mathcal{A}_K$ such that $cJ \subseteq \mathcal{A}_K$.

4.5. Lemma (i) Every ideal $I \subseteq \mathcal{A}_K$ is a fractional ideal of $\mathcal{A}_K$.
(ii) An $\mathcal{A}_K$-submodule J is a fractional ideal of $\mathcal{A}_K$ if and only if there is an ideal $I \subseteq \mathcal{A}_K$ and some $0 \neq c \in \mathcal{A}_K$ such that $c^{-1}I = J$.

Proof Exercise. □

Example (i) If $\mathcal{A}_K$ is a PID, then every fractional ideal J of $\mathcal{A}_K$ is of the form $J = \frac{d}{c}\mathcal{A}_K$, where $\frac{d}{c} \in K$. To see this, note that if $c \in \mathcal{A}_K$ is such that $cJ \subseteq \mathcal{A}_K$, then cJ is an ideal of $\mathcal{A}_K$, and hence $cJ = d\mathcal{A}_K$ for some $d \in \mathcal{A}_K$, that is, $J = \frac{d}{c}\mathcal{A}_K$.

Put

$$\mathcal{F}(\mathcal{A}_K) = \left\{ J \subset K \mid J \text{ a fractional ideal of } \mathcal{A}_K \right\},$$

and define the multiplication on $\mathcal{F}(\mathcal{A}_K)$ by setting

$$J_1 J_2 = \left\{ \sum xy \mid x \in J_1, \, y \in J_2 \right\}, \quad J_1, J_2 \in \mathcal{F}(\mathcal{A}_K).$$

The reader will be asked in later exercise 4 to check that this operation is well-defined and it makes $\mathcal{F}(\mathcal{A}_K)$ into a commutative associative semigroup with the identity element $\mathcal{A}_K$.

We now proceed to show that $\mathcal{F}(\mathcal{A}_K)$ is indeed a group.

For each ideal I of $\mathcal{A}_K$, set

$$I^{-1} = \left\{ x \in K \mid xI \subseteq \mathcal{A}_K \right\}.$$

Then it is clear that $\mathcal{A}_K \subseteq I^{-1}$.

4.6. Proposition With notation as above, the following statements hold:
(i) If I is an ideal of $\mathcal{A}_K$ then $I^{-1} \in \mathcal{F}(\mathcal{A}_K)$. Moreover, II^{-1} is an ideal of $\mathcal{A}_K$.

(ii) If M is a maximal ideal of $\mathcal{A}_K$, then $\mathcal{A}_K \neq M^{-1}$. Consequently, $\mathcal{A}_K \neq I^{-1}$ for any proper ideal I of $\mathcal{A}_K$.
(iii) Let I be an ideal of $\mathcal{A}_K$, $I \neq \{0\}$. If $SI \subseteq I$ for some subset $S \subseteq K$, then $S \subseteq \mathcal{A}_K$.
(iv) If M is a maximal ideal of $\mathcal{A}_K$, then $MM^{-1} = \mathcal{A}_K$.

Proof (i) If $0 \neq c \in I$ then $cT^{-1} \subseteq \mathcal{A}_K$. So $I^{-1} \in \mathcal{F}(\mathcal{A}_K)$. That II^{-1} is an ideal of $\mathcal{A}_K$ may be verified directly.
(ii) Let $0 \neq a \in M$ and $\langle a \rangle$ the ideal of $\mathcal{A}_K$ generated by a. By Proposition 4.3, $P_1 \cdots P_r \subseteq \langle a \rangle \subseteq M$ for some prime ideals $P_1, ..., P_r$. Since M is a prime ideal, $M = P_i$ for some i by Theorem 4.2. Assume $P_i = P_1$ without loss of generality. Choosing r to be smallest, we have $P_2 \cdots P_r \not\subseteq \langle a \rangle$. Taking $b \in P_2 \cdots P_r$, $b \notin \langle a \rangle$, we see that

$$bM = bP_1 \subseteq \langle a \rangle = a\mathcal{A}_K \text{ implies } a^{-1}b \in M^{-1},$$
$$b \notin \langle a \rangle = a\mathcal{A}_K \text{ implies } a^{-1}b \notin \mathcal{A}_K.$$

Thus, if I is a proper ideal of $\mathcal{A}_K$, then $I \subseteq M$ for some maximal ideal M. It follows that $M^{-1} \subseteq I^{-1}$ and $I^{-1} \neq \mathcal{A}_K$.
(iii) Since $\mathcal{A}_K$ is Noetherian, assume $I = \sum_{i=1}^{m} \mathcal{A}_K v_i$, $v_i \in I$. Let $s \in S$. Then $SI \subseteq I$ yields

$$sv_i = a_{i1}v_1 + a_{i2}v_2 + \cdots + a_{im}v_m, \ a_{ij} \in \mathcal{A}_K, \ i = 1, ..., m.$$

This means that the system of linear equations

$$\begin{cases} (a_{11} - s)v_1 + a_{12}v_2 + \cdots + a_{1m}v_m = 0 \\ a_{21}v_1 + (a_{22} - s)v_2 + \cdots + a_{2m}v_m = 0 \\ \vdots \\ a_{m1}v_1 + a_{m2}v_2 + \cdots + (a_{mm} - s)v_m = 0 \end{cases}$$

has a nonzero solution $(v_1, ..., v_m)$. Write A for the coefficient matrix of the system. Working over the field K, A is singular and hence $\det(A) = 0$. This shows that s is a zero of a monic polynomial in $\mathcal{A}_K[x]$. Thus, $s \in \mathcal{A}_K$ by Lemma 1.1.
(iv) By (i), MM^{-1} is an ideal of $\mathcal{A}_K$. Since $\mathcal{A}_K \subset M^{-1}$, $M \subseteq MM^{-1} \subseteq \mathcal{A}_K$. The maximality of M entails $M = MM^{-1}$ or $MM^{-1} = \mathcal{A}_K$. If $M = MM^{-1}$, then (iii) implies $M^{-1} \subseteq \mathcal{A}_K$, contradicting (ii). It follows that $MM^{-1} = \mathcal{A}_K$. □

4.7. Theorem With notation as before, the following hold:

(i) If I is a nonzero ideal of $\mathcal{A}_K$, then $II^{-1} = \mathcal{A}_K$.
(ii) Every nonzero fractional ideal $J \in \mathcal{F}(\mathcal{A}_K)$ has an inverse $J^{-1} \in \mathcal{F}(\mathcal{A}_K)$, i.e., $JJ^{-1} = \mathcal{A}_K$.
Therefore, $\mathcal{F}(\mathcal{A}_K)$ is a commutative multiplicative group.

Proof (i) Suppose that $II^{-1} \neq \mathcal{A}_K$, and that I is maximal among all such ideals (with respect to $\subseteq$). Let M be a maximal ideal of $\mathcal{A}_K$ containing I. Then
$$\mathcal{A}_K \subset M^{-1} \subseteq I^{-1} \text{ implies } I \subseteq IM^{-1} \subseteq II^{-1} \subset \mathcal{A}_K.$$
By Proposition 4.6(ii–iii), $I \neq IM^{-1}$. Note that IM^{-1} is an ideal of $\mathcal{A}_K$. By the choice of I,
$$(IM^{-1})(IM^{-1})^{-1} = \mathcal{A}_K \text{ implies } M^{-1}(IM^{-1})^{-1} \subseteq I^{-1}$$
and this shows that
$$\mathcal{A}_K = (IM^{-1})(IM^{-1})^{-1} \subseteq II^{-1} \subseteq \mathcal{A}_K,$$
that is, $II^{-1} = \mathcal{A}_K$, a contradiction. It follows that we must have $II^{-1} = \mathcal{A}_K$.
(ii) This follows from part (i) and Lemma 4.5(ii). □

Example (ii) Let $K = \mathbb{Q}(\sqrt{-5})$. Then $\mathcal{A}_K = \mathbb{Z}[\sqrt{-5}]$. The ideal $I = \langle 2, 1 - \sqrt{-5} \rangle$ of $\mathcal{A}_K$ has inverse
$$I^{-1} = \left\{ \frac{a}{2} + \frac{c}{2}\sqrt{-5} = x \in K \ \bigg| \ 2 | (a - c) \right\}.$$

We are ready to reach the unique factorization theorem of ideals in $\mathcal{A}_K$.

4.8. Theorem Every nonzero ideal of $\mathcal{A}_K$ can be written as a product of finitely many prime ideals, uniquely up to the order of factors.

Proof If there is some nonzero ideal I of $\mathcal{A}_K$ which cannot be written as a product of finitely many prime ideals, we may assume I is maximal among all such nonzero ideals (with respect to $\subseteq$). Then I is not a prime and it is properly contained in some maximal ideal M. Thus, $M^{-1} \subseteq I^{-1}$, $I \subset IM^{-1} \subseteq \mathcal{A}_K$, and by Proposition 4.6(ii–iii), $I \neq IM^{-1}$. Since IM^{-1} is an ideal of $\mathcal{A}_K$, by the choice of I we have $IM^{-1} = P_1 \cdots P_r$ for some prime ideals $P_1, ..., P_r$. But this yields $I = MP_1 \cdots P_r$, a contradiction.

Therefore, every nonzero ideal of $\mathcal{A}_K$ can be written as a product of finitely many prime ideals.

Now if $I = P_1 \cdots P_r = Q_1 \cdots Q_s = J$ for prime ideals $P_1, ..., P_r, Q_1, ..., Q_s$, then $I \subseteq P_1$ implies $Q_i \subseteq P_1$ for some Q_i. So $P_1 = Q_i$ by their maximality. Assuming $Q_i = Q_1$ and multiplying I and J by P_1^{-1}, we have $P_2 \cdots P_r = Q_2 \cdots Q_s$ and the proof of uniqueness may be done by an inductive demonstration. □

Example (iii) In $\mathcal{A}_K = \mathbb{Z}[\sqrt{-5}] \subset K = \mathbb{Q}(\sqrt{-5})$, we know that $6 = 2 \times 3 = (1+\sqrt{-5})(1-\sqrt{-5})$. But we have a unique factorization of $\langle 6 \rangle$ into prime ideals:

$$\langle 6 \rangle = \langle 2, 1+\sqrt{-5} \rangle^2 \langle 3, 1+\sqrt{-5} \rangle \langle 3, 1-\sqrt{-5} \rangle.$$

Moreover, in $\mathcal{A}_K = \mathbb{Z}[\sqrt{-17}] \subset K = \mathbb{Q}(\sqrt{-17})$, we know that $18 = 2 \times 3 \times 3 = (1+\sqrt{-17})(1-\sqrt{-17})$. But we have a unique factorization of $\langle 18 \rangle$ into prime ideals:

$$\langle 18 \rangle = \langle 1+\sqrt{-17} \rangle \langle 1-\sqrt{-17} \rangle.$$

The reader is asked to check that each ideal used in the above factorizations is prime.

Finally, with the help of Theorem 4.8 we are ready to see how far is $\mathcal{A}_K$ from being a UFD.

4.9. Theorem $\mathcal{A}_K$ is a UFD if and only if it is a PID.

Proof The "if" part is known.

Now suppose $\mathcal{A}_K$ is a UFD. In view of Theorem 4.8 it is sufficient to show that every prime ideal of $\mathcal{A}_K$ is principal. If P is a nonzero prime ideal of $\mathcal{A}_K$, then by Lemma 4.1 we may pick up some integer $0 \neq m \in P$ and consider its unique factorization into irreducible elements in $\mathcal{A}_K$:

$$m = \gamma_1 \gamma_2 \cdots \gamma_s.$$

Then some $\gamma_i \in P$ or $\langle \gamma_i \rangle \subseteq P$. But each γ_i is a prime because $\mathcal{A}_K$ is a UFD (Chapter 1 Theorem 2.9). It follows that $\langle \gamma_i \rangle$ is maximal and hence $\langle \gamma_i \rangle = P$. □

The next theorem will show that $\mathcal{A}_K$ is only one pace away from being a PID (just as indicated in the diagram given in the preface of this book!).

We demonstrate this result through several technical preliminaries.

4.10. Lemma (i) If I and J are ideals of $\mathcal{A}_K$ such that $I + J = \mathcal{A}_K$, then $I \cap J = IJ$.
(ii) Let $P_1, ..., P_r$ be distinct prime ideals of $\mathcal{A}_K$, then $P_1 \cdots P_r = \cap_{i=1}^r P_i$.

Proof Exercise. □

4.11. Proposition (i) Let I be a nonzero ideal of $\mathcal{A}_K$ and P a prime ideal (hence maximal) of $\mathcal{A}_K$. Then for any integer $s \geq 1$ there exists $b \in I$ such that $bI^{-1} + P^s = \mathcal{A}_K$.
(ii) Let I, J be nonzero ideals of $\mathcal{A}_K$, and $J = P_1^{s_1} \cdots P_r^{s_r}$ the unique prime factorization of J in $\mathcal{A}_K$, where $r \geq 1$, $s_i \geq 1$, $i = 1, ..., r$. If there is some $0 \neq b \in \cap_{i=1}^r (I - IP_i)$, then $bI^{-1} + J = \mathcal{A}_K$.
(iii) Let I and $J = P_1^{s_1} \cdots P_r^{s_r}$ be as in part (ii), then there is some $0 \neq b \in \cap_{i=1}^r (I - IP_i)$.

Proof (i) Note that $IP \subset I$ but $IP \neq I$ (otherwise $P = I^{-1}(IP) = I^{-1}I = \mathcal{A}_K$). Let $0 \neq b \in I - IP$. Then

(1) $$bI^{-1} \not\subseteq P.$$

If $bI^{-1} = \mathcal{A}_K$, then $bI^{-1} + P^s = \mathcal{A}_K$. If $bI^{-1} \neq \mathcal{A}_K$, then by the above (1) we have,

(2) $$bI^{-1} + P = \mathcal{A}_K.$$

It follows that

$$bI^{-1}P + P^2 = P, \text{ and by substitution in (2), } bI^{-1} + P^2 = \mathcal{A}_K.$$

After repeating the same process for $s - 2$ times we arrive at

$$bI^{-1} + P^s = \mathcal{A}_K.$$

(ii) By the assumption, $bI^{-1}P_i \not\subseteq P_i$, $i = 1, ..., r$. If $bI^{-1} = \mathcal{A}_K$, then $bI^{-1} + J = \mathcal{A}_K$. If $bI^{-1} \neq \mathcal{A}_K$, then a similar argumentation as in the proof of part (i) shows that

$$bI^{-1} + P_i^{s_i} = \mathcal{A}_K, \ i = 1, ..., r.$$

It follows that

$$bI^{-1}P_j^{s_j} + P_i^{s_i}P_j^{s_j} = P_j^{s_j}, \ i \neq j.$$

After substitution step by step, we arrive at
$$bI^{-1} + J = bI^{-1} + P_1^{s_1} \cdots P_r^{s_r} = \mathcal{A}_K.$$

(iii) For each $i = 1, ..., r$, let $I_i = IP_1 \cdots P_r P_i^{-1}$. Then

(3) $$I_i = IP_1 \cdots P_{i-1}P_{i+1} \cdots P_r \subseteq IP_j, \quad j \neq i.$$

Since $I_i \neq I_i P_i$ (otherwise $P_i = \mathcal{A}_K$), there is $b_i \in I_i - I_i P_i$, $i = 1, ..., r$. Set
$$b = b_1 + \cdots + b_r.$$

Then $b \in I$ and we claim that $b \notin IP_i$, $i = 1, ..., r$. To see this, suppose that $b \in IP_j$ for some j. Then by the construction of b and (3) above, $b_j \in IP_j$ would hold. But we know that
$$b_j \notin I_j P_j = \left(IP_1 \cdots P_r P_j^{-1}\right) P_j = IP_1 \cdots P_r.$$

By Lemma 4.10(ii), this yields

(4) $$b_j I^{-1} \not\subseteq P_1 \cdots P_r = \bigcap_{i=1}^r P_i.$$

Moreover, $b_j \in I_j$ implies

(5) $$b_j I^{-1} \subseteq I^{-1} I_j = P_1 \cdots P_{j-1} P_{j+1} \cdots P_r.$$

Thus, by (5) above, $b_j \in IP_j$, or equivalently, $b_j I^{-1} \subseteq P_j$ would yield
$$b_j I^{-1} \subseteq P_j \cap P_1 \cdots P_{j-1} P_{j+1} \cdots P_r = \bigcap_{i=1}^r P_i,$$

contradicting the foregoing (4). Therefore, we must have $b \in \cap_{i=1}^r (I - IP_i)$. □

4.12. Theorem Let I be a nonzero ideal of $\mathcal{A}_K$, $0 \neq a \in I$. Then there is some $b \in I$ such that $I = a\mathcal{A}_K + b\mathcal{A}_K$.

Proof For $0 \neq a \in I$, set $J = aI^{-1}$. Then by Proposition 4.11(ii–iii), there is $b \in I$ such that $bI^{-1} + aI^{-1} = \mathcal{A}_K$, and it follows that $I = b\mathcal{A}_k + a\mathcal{A}_K$. □

To measure the extent to which ideals in $\mathcal{A}_K$ are not principal, another topic in algebraic number theory is to study the class-group of K, that is the quotient group of $\mathcal{F}(\mathcal{A}_K)$ by the (normal) subgroup of principal fractional ideals.

From $\mathcal{A}_K$ to Dedekind domains

Historically in the literature, if a normal Noetherian domain R also has the property that every nonzero prime ideal is maximal, then R is called a *Dedekind domain* after the mathematicians, such as Kummer, Dedekind in the 19th century and Noether in the 1930s, who developed the ring theoretic approach to the study of $\mathcal{A}_K$. With no modification, all results about $\mathcal{A}_K$ given after Theorem 4.2 in this section hold for a Dedekind domain R, in particular,
- every nonzero ideal of R has a unique factorization into primes;
- every ideal of R can be generated by two elements; and
- R is a UFD if and only if it is a PID.

Moreover, there is also the class-group theory associated to the ideals of a Dedekind domain that has an intimate connection with the divisor class-group of an algebraic curve in algebraic geometry.

Exercises

1. Let $\alpha \in \mathcal{A}_K$ be an irreducible element but not a prime. If $\langle \alpha \rangle = P_1 \cdots P_s$ is the unique factorization of $\langle \alpha \rangle$ into prime ideals in $\mathcal{A}_K$, show that none of P_i is principal.
2. Show that $\langle 3, 1 + \sqrt{-5} \rangle$ is a prime ideal in $\mathcal{A}_K = \mathbb{Z}[\sqrt{-5}]$, and that $\langle 1 + \sqrt{-17} \rangle$ is a prime ideal in $\mathcal{A}_K = \mathbb{Z}[\sqrt{-17}]$.
3. Complete the proof of Lemma 4.5. (Hint: To reach (ii), note that if $cJ \subset \mathcal{A}_K$ for some $0 \neq c \in \mathcal{A}_K$ then cJ is an ideal of $\mathcal{A}_K$.)
4. Show that the operation on $\mathcal{F}(\mathcal{A}_K)$ is well-defined and it makes $\mathcal{F}(\mathcal{A}_K)$ into a commutative associative semigroup with the identity element $\mathcal{A}_K$.
5. Complete the proof of Lemma 4.10.

Chapter 5
Algebraic Geometry

This chapter demonstrates how Noetherian rings, field extensions, local rings, DVRs, normalization and localization stem naturally from algebraic geometry.

Throughout the text $K[x_1, ..., x_n]$ denotes the polynomial ring in $x_1, ..., x_n$ over a field K. If $n = 2$ we use $K[x, y]$ instead of $K[x_1, x_2]$, and similarly we use $K[x, y, z]$ in place of $K[x_1, x_2, x_3]$. Moreover, let

$$\mathbf{A}^n = \mathbf{A}^n_K = \left\{ P = (a_1, ..., a_n) \mid a_1, ..., a_n \in K \right\}$$

stand for the n-dimensional *affine space* (or affine n-space) over K. An element $P \in \mathbf{A}^n$ is called a *point*, and if $P = (a_1, ..., a_n)$ then a_i is called the *ith coordinate* of P.

It is clear that in the case where $K = \mathbb{R}$, $\mathbf{A}^1 = \mathbb{R}$ is the real affine line, $\mathbf{A}^2 = \mathbb{R}^2$ is the real affine plane, and $\mathbf{A}^3 = \mathbb{R}^3$ is the real affine space.

1. Finite Field Extension and Nullstellensatz

Let $f \in K[x_1, ..., x_n]$. Then f defines a function

$$\phi_f : \quad \mathbf{A}^n = \mathbf{A}^n_K \longrightarrow K$$
$$P = (a_1, ..., a_n) \mapsto f(P) = f(a_1, ..., a_n)$$

If $f = c \in K$ is a constant, then $\phi_f(P) = c$ defines a constant function. If $f, g \in K[x_1, ..., x_n]$, then, as usual functions, $(\phi_f + \phi_g)(P) = \phi_f(P) + \phi_g(P)$ and $(\phi_f \phi_g)(P) = \phi_f(P)\phi_g(P)$.

1.1. Definition Let T be a subset of $K[x_1, ..., x_n]$, and

$$\mathbf{V}(T) = \left\{ P \in \mathbf{A}^n \mid \phi_f(P) = f(P) = 0, \text{ for all } f \in T \right\},$$

that is, $\mathbf{V}(T)$ is the set of common zeros of all polynomials in T, or equivalently the set of solutions of the system of polynomial equations

$$f_i = 0, \quad f_j \in T.$$

We call $\mathbf{V}(T)$ an *affine algebraic set* (or just an *algebraic set*) of the affine n-space $\mathbf{A}^n = \mathbf{A}^n_K$. If $T = \{f_1, ..., f_s\}$ is a finite subset, we also write $\mathbf{V}(T) = \mathbf{V}(f_1, ..., f_s)$.

If $f \in K[x_1, ..., x_n]$ is not a constant, the algebraic set $\mathbf{V}(f)$ is called the *hypersurface* defined by f; if $\deg(f) = 1$, $\mathbf{V}(f)$ is called a *hyperplane* in $\mathbf{A}^n$. A hypersurface in $\mathbf{A}^2$ is called a *plane curve*; and a hyperplane in $\mathbf{A}^2$ is called a *line*.

Example (i) $\mathbf{A}^n$ and $\emptyset$ are algebraic sets (indeed $\mathbf{A}^n = \mathbf{V}(0)$, $\emptyset = \mathbf{V}(1) = \mathbf{V}(c)$ for any $c \in K^\times$).

(ii) Conics in $\mathbb{R}^2$ defined by quadratic polynomials in $\mathbb{R}[x, y]$ are plane curves; while $\mathcal{C} = \mathbf{V}(y^2 - x^3) \subset \mathbf{A}^2_\mathbb{R}$ is known as the *cuspidal curve* and $\mathcal{C} = \mathbf{V}(y^2 - x^3 - x^2) \subset \mathbf{A}^2_\mathbb{R}$ is known as the *nodal curve*. Moreover, $V = \mathbf{V}(y - x^2, z - x^3) \subset \mathbf{A}^3_\mathbb{R}$ is known as the *twisted cubic* in $\mathbb{R}^3$.

1.2. Proposition (i) Let T be a nonempty subset of $K[x_1, ..., x_n]$ and $I = \langle T \rangle$ the ideal generated by T in $K[x_1, ..., x_n]$. Then $\mathbf{V}(T) = \mathbf{V}(I) = \mathbf{V}(f_1, ..., f_s)$ for finitely many $f_1, ..., f_s \in T$.
(ii) If $T_1 \subseteq T_2$ are subsets of $K[x_1, ..., x_n]$, then $\mathbf{V}(T_1) \supseteq \mathbf{V}(T_2)$.
(iii) If Y_1, Y_2 are algebraic sets, then so is $Y_1 \cup Y_2$. Consequently, the union of any finitely many algebraic sets is an algebraic set.
(iv) If $\{Y_i\}_{i \in J}$ is a family of algebraic sets, then $\cap_{i \in J} Y_i$ is an algebraic set.

Proof The existence of $f_1, ..., f_s$ follows from the fact that $K[x_1, ..., x_n]$ is Noetherian (Chapter 1 Theorem 1.3). All other assertions may be directly verified by means of part (i) and Definition 1.1. □

Proposition 1.2(i) establishes the initial connection between ideals and algebraic sets that makes the flexibility of finding easier defining equations of a given system of polynomial equations.

Example (iii) Let $V = \mathbf{V}(f_1, f_2)$ and $I = \langle f_1, f_2 \rangle$, where $f_1 = x^2 - xy - y^2 + z^2$, $f_2 = x^2 - y^2 + z^2 - z \in \mathbb{R}[x, y, z]$. Then since $I = \langle f_1, f_2 \rangle = \langle f_1, f_1 + cf_2 \rangle$ for any nonzero $c \in \mathbb{R}$, it follows that, after setting $c = -1$, $V = \mathbf{V}(f_1, z - xy)$, i.e., every point $P \in V$ is of the form $P = (x, y, xy)$.

(iv) Since for $a_i \in k$, $\mathbf{V}(x_1 - a_1, ..., x_n - a_n) = \{P = (a_1, ..., a_n)\} \subset \mathbf{A}^n$, it follows from Proposition 1.2(iii) that any finite subset of $\mathbf{A}^n$ is an algebraic set.

For ideals in $K[x_1, ..., x_n]$, the operations sum, product, and intersection correspond to operations of algebraic sets.

1.3. Proposition Let I and J be ideals in $K[x_1, ..., x_n]$. The following hold:
(i) $\mathbf{V}(I + J) = \mathbf{V}(I) \cap \mathbf{V}(J)$.
(ii) $\mathbf{V}(I \cdot J) = \mathbf{V}(I) \cup \mathbf{V}(J)$.
(iii) $\mathbf{V}(I \cap J) = \mathbf{V}(I) \cup \mathbf{V}(J)$.

Proof Exercise. □

Now, if we start with a subset $Y \subseteq \mathbf{A}^n$, then Y may be associated to a set of polynomials

$$\mathbf{I}(Y) = \left\{ f \in K[x_1, ..., x_n] \;\middle|\; f(P) = 0 \text{ for all } P \in Y \right\}.$$

It is easily seen that $\mathbf{I}(Y)$ is an ideal of $K[x_1, ..., x_n]$.

$\mathbf{I}(Y)$ is called the *ideal* of Y in $K[x_1, ..., x_n]$. As we will see from now on, it is this ideal that makes the essential connection between algebraic structure theory and the geometry of algebraic sets.

1.4. Proposition (i) If J is an ideal of $K[x_1, ..., x_n]$, then $J \subseteq \mathbf{I}(\mathbf{V}(J))$, and $\mathbf{V}(\mathbf{I}(\mathbf{V}(J))) = \mathbf{V}(J)$.
(ii) For any subset $Y \subseteq \mathbf{A}^n$, $\mathbf{V}(\mathbf{I}(Y)) = \overline{Y}$ is the smallest affine algebraic set containing Y.
(iii) For any two subsets Y_1, Y_2 of $\mathbf{A}^n$, we have $\mathbf{I}(Y_1 \cup Y_2) = \mathbf{I}(Y_1) \cap \mathbf{I}(Y_2)$.
(iv) Let V and W be affine algebraic sets in $\mathbf{A}^n$. Then:
  (a) $V \subseteq W$ if and only if $\mathbf{I}(V) \supseteq \mathbf{I}(W)$.
  (b) $V = W$ if and only if $\mathbf{I}(V) = \mathbf{I}(W)$.

Proof Exercise. □

Example (v) If K is infinite then $\mathbf{I}(\mathbf{A}^n) = \{0\}$ (see later exercise 2). If $P = (a_1, a_2, ..., a_n) \in \mathbf{A}^n$, then $\mathbf{I}(\{P\}) = \langle x_1 - a_1, x_2 - a_2, ..., x_n - a_n \rangle$ which is a maximal ideal of $K[x_1, ..., x_n]$ (Chapter 2 section 1, Example (iii)).

(vi) For $V = \mathbf{V}(y - x^2) \subset \mathbf{A}^2_{\mathbb{R}}$, $\mathbf{I}(V) = \langle y - x^2 \rangle$. This may be verified by noticing that any monomial $x^\alpha y^\beta$ in $\mathbb{R}[x, y]$ can be written as $\sum_i (y - x^2) h_i + x^{\alpha + 2\beta}$ where $h_i \in \mathbb{R}[x, y]$, and any point $P \in V$ is of the form $P = (x, x^2)$, $x \in \mathbb{R}$.

(vii) Note that $\langle x^2, y^2 \rangle$ is properly contained in $\mathbf{I}(\mathbf{V}(x^2, y^2)) = \langle x, y \rangle$. So in general J is not necessarily equal to $\mathbf{I}(\mathbf{V}(J))$.

So far we have built two mappings

$$\{\text{ideals}\} \xrightarrow{\mathbf{V}} \{\text{affine algebraic sets}\}$$

$$J \mapsto \mathbf{V}(J)$$

and

$$\{\text{affine algebraic sets}\} \xrightarrow{\mathbf{I}} \{\text{ideals}\}$$

$$V \mapsto \mathbf{I}(V)$$

Note that by Proposition 1.4 the second mapping is injective. The first mapping, however, is not necessarily injective: *different ideals can define the same algebraic sets*. For example, the ideals $\langle x \rangle$ and $\langle x^2 \rangle$ are different in $K[x]$, where K is an arbitrary field, but $\mathbf{V}(x) = \mathbf{V}(x^2) = \{0\}$ in $\mathbf{A}^1_K$. Also consider the ideals of $\mathbb{R}[x]$:

$$J_1 = \langle 1 \rangle = \mathbb{R}[x], \quad J_2 = \langle 1 + x^2 \rangle, \quad J_3 = \langle 1 + x^2 + x^4 \rangle.$$

We see that $J_i \neq J_k$ if $i \neq k$. But $\mathbf{V}(I_1) = \mathbf{V}(I_2) = \mathbf{V}(I_3) = \emptyset$. Similarly, $\langle x^3 - 1 \rangle \neq \langle xy^2 - y^2 + x - 1 \rangle$ in $\mathbb{R}[x, y]$, but both ideals correspond to the same algebraic set $\{(1, y) \mid y \in \mathbb{R}\}$.

The wonderful thing is that if the ground field K is algebraically closed then the Hilbert's Nullstellensatz solves the problem entirely.

1.5. Theorem (weak Nullstellensatz) If the field K is algebraically closed, then there is a natural one-to-one and onto correspondence

$$\mathbf{A}^n \longrightarrow \text{m-Spec}\, K[x_1, ..., x_n]$$
$$(a_1, ..., a_n) \mapsto \langle x_1 - a_1, ..., x_n - a_n \rangle$$

Proof For each point $P = (a_1, ..., a_n) \in \mathbf{A}^n$, that $\langle x_1 - a_1, ..., x_n - a_n \rangle$ is a maximal ideal of $K[x_1, ..., x_n]$ follows from Chapter 2 (section 1, Example (iii)). Conversely, let $M \in$ m-Spec$K[x_1, ..., x_n]$ and write $K[\bar{x}_1, ..., \bar{x}_n] = K[x_1, ..., x_n]/M$, where each $\bar{x}_i$ is the image of x_i in $K[x_1, ..., x_n]/M$. Then the natural K-algebra homomorphisms

$$K \hookrightarrow K[x_1, ..., x_n] \xrightarrow{\pi} K[\bar{x}_1, ..., \bar{x}_n]$$

yield a finitely generated field extension $K \subset K[\bar{x}_1, ..., \bar{x}_n]$ (here K and $\pi(K)$ are identified). By Chapter 3 Theorem 2.5, $K[\bar{x}_1, ..., \bar{x}_n]$ is algebraic over K. But K is algebraically closed by the assumption. Hence $K = K[\bar{x}_1, ..., \bar{x}_n]$. Writing $\bar{x}_i = b_i \in K$, $i = 1, ..., n$, then π is just given by

$$f(x_1, ..., x_n) \mapsto f(b_1, ..., b_n)$$

and it follows that Ker$\pi = M = \langle x_1 - b_1, ..., x_n - b_n \rangle$ by Chapter 2 (section 1, Example (iii)). That the correspondence is one-to-one and onto is easily seen now. □

1.6. Theorem (Nullstellensatz) Let K be an algebraically closed field, and let J be an ideal of $K[x_1, ..., x_n]$.
(i) If $J \neq K[x_1, ..., x_n]$, then $\mathbf{V}(J) \neq \emptyset$.
(ii) $\mathbf{I}(\mathbf{V}(J)) = \sqrt{J}$, where $\sqrt{J} = \{f \in K[x_1, ..., x_n] \mid f^m \in J \text{ for some } m \geq 1\}$ is the radical of J (Chapter 2 section 1, exercise 2).

Proof (i) By the assumption and Theorem 1.5,

$$J \subseteq M = \langle x_1 - a_1, ..., x_n - a_n \rangle$$

for some $P = (a_1, ..., a_n) \in \mathbf{A}^n$. Thus, $P \in \mathbf{V}(M) \subseteq \mathbf{V}(J)$.
(ii) By the definition of $\sqrt{J}$, the inclusion $\sqrt{J} \subseteq \mathbf{I}(\mathbf{V}(J))$ is clear. To reach the inclusion $\sqrt{J} \supseteq \mathbf{I}(\mathbf{V}(J))$, let $J = \langle f_1, ..., f_s \rangle$, and $0 \neq f \in \mathbf{I}(\mathbf{V}(J))$. Then $S = \{1, f, f^2, ..., f^m, ...\}$ is a multiplicative set of $K[x_1, ..., x_n]$. Consider the ring of fractions $K[x_1, ..., x_n]_S$. By Chapter 2 (section 3, Example (ii)),

$$K[x_1, ..., x_n]_S \cong \frac{K[x_1, ..., x_n][y]}{\langle yf - 1 \rangle}.$$

Note that $K[x_1, ..., x_n]_S \neq \{0\}$. So $\langle yf - 1 \rangle \neq K[x_1, ..., x_n][y]$. But we claim

$$J^* = \langle J, \; fy - 1 \rangle = K[x_1, ..., x_n, y],$$

for, Theorem 1.5 asserts that there is no maximal ideal of $K[x_1, ..., x_n, y]$ containing J^*. Thus,

$$(1) \quad 1 = \sum_{i=1}^{s} p_i(x_1, ..., x_n, y) f_i + q(x_1, ..., x_n, y)(1 - yf)$$

for some polynomials $p_i, q \in K[x_1, ..., x_n, y]$. Now, going back to $K[x_1, ..., x_n]_S$ via $y \mapsto \frac{1}{f}$ (see Chapter 2 (section 3, Example (ii))), the relation (1) above implies that

$$(2) \quad 1 = \sum_{i=1}^{s} p_i\left(x_1, ..., x_n, \frac{1}{f}\right) f_i.$$

After clearing all denominators of Eq. (2) by a suitable power f^m, we obtain

$$(3) \quad f^m = \sum_{i=1}^{s} g_i f_i,$$

for some polynomials $g_i \in K[x_1, ..., x_n]$, as desired. □

If an ideal $J \subset K[x_1, ..., x_n]$ has the property that $J = \sqrt{J}$, then J is called a *radical ideal*. It is clear that $\mathbf{I}(\mathbf{V}(J))$ is radical, and every prime ideal is radical. But not every ideal is radical, as easily illustrated by the ideal $\langle x^2, y^2 \rangle \subset \mathbb{R}[x, y]$ (also see later exercise 9). Working with radical ideals, the Nullstellensatz establishes the following perfect correspondences.

1.7. Theorem If K is algebraically closed, then the mappings

$$\{\text{affine algebraic sets}\} \xrightarrow{\mathbf{I}} \{\text{radical ideals}\}$$

and

$$\{\text{radical ideals}\} \xrightarrow{\mathbf{V}} \{\text{affine algebraic sets}\}$$

are inclusion-reversing bijections which are inverses to each other.

□

Theorem 1.6 answered when an algebraic set is nonempty, provided the ground field is algebraically closed. The next proposition answers how to algebraically recognize the finiteness of an algebraic set $\mathbf{V}(J)$.

1.8. Theorem Let J be an ideal of $K[x_1, ..., x_n]$, $V = \mathbf{V}(J)$, and $R = K[x_1, ..., x_n]/J$.
(i) If $\dim_K R < \infty$, then V is finite.
(ii) If K is algebraically closed and V is finite, then $\dim_K R < \infty$.

Proof (i) Let $\dim_K R = m$ and $\bar{x}_i$ the image of x_i in R, $i = 1, ..., n$. Then
$$\lambda_m \bar{x}_i^m + \lambda_{m-1} \bar{x}_i^{m-1} + \cdots + \lambda_1 \bar{x}_i + \lambda_0 = 0, \text{ for some } \lambda_i \in K.$$
This implies the one-variable polynomial $\sum_{j=0}^m x_i^j = f(x_i) \in J$. Hence, $f(x_i)$ vanishes at every point of V. Since $f(x_i)$ can have only finitely many zeros in K, it follows that the points of V have only finitely distinct ith coordinates, $i = 1, ..., n$. This shows that V is finite.
(ii) Note that K is algebraically closed. If $V = \emptyset$, then $1 \in J$ by Nullstelensatz, and hence $\dim_K R = 0$. Suppose $V \neq \emptyset$, and let a_{ij} be the distinct ith coordinates of all s points in V, $i = 1, ..., n$, $j = 1, ..., s$. Define the one-variable polynomials
$$f_i = \prod_{j=1}^s (x_i - a_{ij}), \ i = 1, ..., n.$$
Then f_i vanishes at every point of V. By Nullstellensatz, $f_i^{m_i} \in J$ for some $m_i \geq 1$. Noticing that f_i is a monic polynomial in one variable, a formal division on elements of $K[x_1, ..., x_{i-1}, x_{i+1}, ..., x_n][x_i]$ by f_i, $i = 1, ..., n$, shows that $\dim_K R \leq \infty$.

Exercises

1. Prove Propositions 1.3–1.4.
2. Let f be a nonconstant polynomial in $K[x_1, ..., x_n]$, where K is algebraically closed. Show that $\mathbf{A}^n - \mathbf{V}(f)$ is infinite if $n \geq 1$, and $\mathbf{V}(f)$ is infinite if $n \geq 2$. Conclude that the complement of any algebraic set is infinite. (Hint: Note that K is infinite by Chapter 1 (section 3, exercise 5). f defines the zero function on K^n if and only if $f = 0$ in $K[x_1, ..., x_n]$.)
3. Show that if K is infinite then $\mathbf{I}(\mathbf{A}^n) = \{0\}$. Is this still true if K is finite?
4. Consider the plane curve $C = \mathbf{V}(f) \subset \mathbf{A}_K^2$ where f is a polynomial of degree n in $K[x, y]$. If L is a line in $\mathbf{A}_K^2$ and $L \not\subset C$, show that $L \cap C$ is a finite set of no more than n points. (Hint: Suppose $L = \mathbf{V}(y - (ax + b))$, and consider $f(x, ax + b) \in K[x]$.)
5. Let $V = \mathbf{V}(f, g) \subset \mathbf{A}_\mathbb{C}^2$, where $f = x^2 - y + 1$, $g = y + x^2 - 5$.

(a) Show that $V = \mathbf{V}(x^2 - y + 1, x^2 - 2)$.
(b) Use part (a) to determine V.

6. This exercise is to demonstrate that factorization of polynomials may help to determine an algebraic set.
   (a) Show that if $g \in K[x_1, ..., x_n]$ factors as $g = g_1 g_2$, where K is a field, then for any f, $\mathbf{V}(f,g) = \mathbf{V}(f, g_1) \cup \mathbf{V}(f, g_2)$.
   (b) Show that in $\mathbf{A}_\mathbb{R}^3$, $\mathbf{V}(y - x^2, xz - y^3) = \mathbf{V}(y - x^2, xz - x^4)$.
   (c) Use part (a) to determine the algebraic set in part (b).

7. For $V = \mathbf{V}(xy - 1) \subset \mathbf{A}_\mathbb{R}^2$, show that $\mathbf{I}(V) = \langle xy - 1 \rangle$.

8. For the twisted cubic $V = \mathbf{V}(y - x^2, z - x^3) \subset \mathbf{A}_\mathbb{R}^3$, show that $\mathbf{I}(V) = \langle y - x^2, z - x^3 \rangle$. (Hint: Any monomial $x^\alpha y^\beta z^\gamma$ in $\mathbb{R}[x, y, z]$ can be written as $\sum_i (y - x^2) h_i + \sum_j (z - x^3) g_j + x^{\alpha + 2\beta + 3\gamma}$, and every point $P \in V$ is of the form $P = (x, x^2, x^3)$.)

9. Let $J = \langle x, y^3 + 1 \rangle \subset \mathbb{R}[x, y]$. Show that $J \neq \sqrt{J}$, and that $\dim_\mathbb{R} \mathbb{R}[x,y]/\sqrt{J}$ is finite.

10. Let $I = \langle x, y \rangle \subset K[x, y]$, where K is a field. Show that $\dim_K (K[x,y]/I^n) = \frac{n(n+1)}{2}$ for all $n \geq 1$.

11. Let J be a proper ideal of $K[x_1, ..., x_n]$. If K is algebraically closed, show that $\sqrt{J} = \mathbf{I}(\mathbf{V}(J)) = \cap M$, where M runs over all maximal ideals containing J.

2. Irreducible V and the Prime $\mathbf{I}(V)$

To study the points of an algebraic set in an "analytic-like" way, we need irreducible algebraic sets and use only "polynomial functions and rational functions" that respect the Zariski topology on the affine n-space $\mathbf{A}^n = \mathbf{A}_K^n$ over a field K (this viewpoint will be clarified in the next section).

Some basic notions and examples on topological space are given in the appendix of this section.

2.1. Definition The *Zariski topology* on $\mathbf{A}^n$ is defined by taking the open subsets to be the complements of affine algebraic sets. For any subset $Y \subset \mathbf{A}^n$, the Zariski topology on Y is the induced topology.

Example (i) Let us consider the Zariski topology on the affine line $\mathbf{A}^1 = K$. Every ideal of $K[x]$ is principal, so every algebraic set is the zero locus of a single polynomial. Thus the algebraic sets in $\mathbf{A}^1$ are just the finite subsets (including the empty set) and the whole space (corresponding to $f = 0$). Consequently the open sets are the empty set and the complements of finite subsets.

(ii) In general, the Zariski topology on $\mathbf{A}^n$ ($n \geq 1$) is not the product topology of the Zariski topology on $\mathbf{A}^1$. For example, if we identify $\mathbf{A}^1 \times \mathbf{A}^1$ with A^2 and consider any curve $C = \mathbf{V}(y - f(x))$ in $\mathbf{A}^2$, where $f(x)$ is a polynomial in $K[x]$ of degree ≥ 1, then from (i) above it is easy to see that the open subset $U = \mathbf{A}^2 - C$ in $\mathbf{A}^2$ cannot be a product of two open subsets of $\mathbf{A}^1$.

2.2. Definition A nonempty subset Y of a topological space X is *irreducible* if it cannot be expressed as the union $Y = Y_1 \cup Y_2$ of two proper subsets, each of which is closed in Y (where Y has the induced topology). The empty subset $\emptyset$ is not considered to be irreducible. If X is not irreducible then it is *reducible*.

Example (iii) If K is infinite, then $\mathbf{A}_K^1$ is irreducible, for its only proper closed subsets are finite sets.

(iv) Similarly, by Proposition 1.3(ii) and (section 1, exercise 3), $\mathbf{A}^n$ is irreducible provided K is infinite.

(v) $V = \mathbf{V}(3x+y-1, y^2-x)$ is reducible in $\mathbf{A}_\mathbb{R}^2$, for V consists of two points. More reducible algebraic sets may be obtained via Proposition 1.3(ii).

The following proposition states two basic properties of irreducible subsets in a topological space. The first one plays a *key* role in algebraic geometry because of previous Example (iv).

2.3. Proposition (i) Any nonempty open subset of an *irreducible space* is dense and irreducible.
(ii) If Y is an irreducible subset of the topological space X, then its closure $\overline{Y}$ in X is also irreducible.

Proof (i) Let U, V be nonempty open subsets of the irreducible space X. Then $U \cap V = \emptyset$ leads to $X = (X - U) \cup (X - V)$, a contradiction. Hence U is dense. If $U = U_1 \cup U_2$ for closed U_1, U_2 in U, then there are proper closed V_1, V_2 in X such that $V_1 \cap U = U_1$, $V_2 \cap U = U_2$. But then $X = (V_1 \cup V_2) \cup (X - U)$, both $V_1 \cup V_2$ and $X - U$ are proper closed subsets.

Once again this is a contradiction.
(ii) Exercise. □

To determine the irreducibility of an algebraic set $V \subset \mathbf{A}^n$ in a purely algebraic way, let us write

$$K[V] = K[x_1, ..., x_n]/\mathbf{I}(V).$$

$K[V]$ is called the *coordinate ring* of V (a remark on this name is given in the beginning of section 3).

2.4. Theorem Let $V \subset \mathbf{A}^n$ be an affine algebraic set. Then V is irreducible if and only if $\mathbf{I}(V)$ is a prime ideal if and only if its coordinate ring $K[V]$ is a domain.

Proof First, assume that V is irreducible and let $fg \in \mathbf{I}(V)$. Set $V_1 = V \cap \mathbf{V}(f)$ and $V_2 = V \cap \mathbf{V}(g)$. Then $fg \in \mathbf{I}(V)$ implies that $V = V_1 \cup V_2$. Since V is irreducible, we have either $V = V_1$ or $V = V_2$. Say the former holds, so that $V = V_1 = V \cap \mathbf{V}(f)$. This implies that f vanishes on V, so that $f \in \mathbf{I}(V)$. Thus, $\mathbf{I}(V)$ is prime.

Next, assume that $\mathbf{I}(V)$ is prime and let $V = V_1 \cup V_2$. Suppose that $V \neq V_1$. We claim that $\mathbf{I}(V) = \mathbf{I}(V_2)$. To prove this, note that $\mathbf{I}(V) \subset \mathbf{I}(V_2)$ since $V_2 \subset V$. For the opposite inclusion, first note that $\mathbf{I}(V)$ is properly contained in $\mathbf{I}(V_1)$ since V_1 is properly contained in V. Thus, we can pick $f \in \mathbf{I}(V_1) - \mathbf{I}(V)$. Now take any $g \in \mathbf{I}(V_2)$. Since $V = V_1 \cup V_2$, it follows that fg vanishes on V, and, hence, $fg \in \mathbf{I}(V)$. But $\mathbf{I}(V)$ is prime, so that f or g lies in $\mathbf{I}(V)$. We know that $f \notin \mathbf{I}(V)$. So, $g \in \mathbf{I}(V)$. This proves $\mathbf{I}(V) = \mathbf{I}(V_2)$, whence $V = V_2$ by Proposition 1.4(iv)(b). Therefore V is an irreducible algebraic set. □

In the "algebra-geometry" dictionary of section 1, algebraic sets in $\mathbf{A}^n$ are associated with ideals in $K[x_1, ..., x_n]$. Note that the ring homomorphism $K[x_1, ..., x_n] \to K[x_1, ..., x_n]/\mathbf{I}(V) = k[V]$ yields the following one-to-one and onto mappings between the given sets:
(a) {ideals of $K[x_1, ..., x_n]$ containing $\mathbf{I}(V)$} and {ideals of $K[V]$}.
(b) {radical ideals of $K[x_1, ..., x_n]$ containing $\mathbf{I}(V)$} and {radical ideals of $K[V]$}.
(c) {prime ideals of $K[x_1, ..., x_n]$ containing $\mathbf{I}(V)$} and {prime ideals of $K[V]$}.
(d) { maximal ideals of $K[x_1, ..., x_n]$ containing $\mathbf{I}(V)$} and {maximal ideals

of $K[V]\}$.
Consequently, the following corollary of Theorem 1.7 is obtained.

2.5. Corollary Let K be an algebraically closed field and let $V \subset \mathbf{A}^n$ be an affine algebraic set.
(i) The correspondences
$$\left\{ \begin{array}{c} \text{affine algebraic subsets} \\ W \subset V \end{array} \right\} \begin{array}{c} \mathbf{I}_V \\ \longrightarrow \\ \longleftarrow \\ \mathbf{V}_V \end{array} \left\{ \begin{array}{c} \text{radical ideals} \\ J \subset K[V] \end{array} \right\}$$
are inclusion-reversing bijections and are inverses to each other, where $\mathbf{I}_V$ denotes taking ideal $\mathbf{I}(W)$ with W as indicated above and $\mathbf{V}_V$ denotes taking algebraic set $\mathbf{V}(J)$ with J as indicated above; we use $\mathbf{I}_V$, respectively $\mathbf{V}_V$, just for the restrictions of $\mathbf{I}$ and $\mathbf{V}$ respectively.
(ii) Under the correspondence given in part (i), irreducible algebraic subsets, respectively points of V, correspond to prime ideals, respectively correspond to maximal ideals of $K[V]$.

Proof Exercise. □

Example (vi) If we plot the twisted cubic $V = \mathbf{V}(y - x^2, z - x^3) \subset \mathbf{A}_{\mathbb{R}}^3$ in a limited space, then an intuitive feeling tells that V is irreducible.

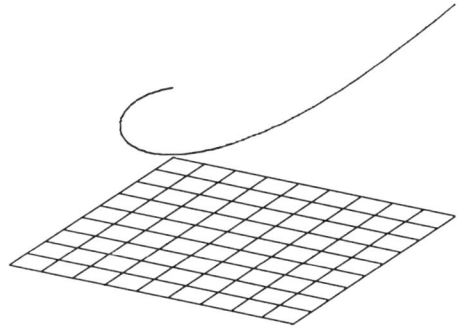

To convince ourselves, however, Theorem 2.4 requires proving that $\mathbf{I}(V)$

is a prime ideal of $\mathbb{R}[x,y,z]$, that is, if $fg \in \mathbf{I}(V)$, we have to show that $f \in \mathbf{I}(V)$ or $g \in \mathbf{I}(V)$.

For any $P = (a,b,c) \in V$, note that setting $\lambda = a$ yields

$$a = \lambda, \quad b = \lambda^2, \quad c = \lambda^3.$$

Conversely, any $\lambda \in \mathbf{A}^1 = \mathbb{R}$ determines a point of V as defined above. Thus, we have $f(\lambda, \lambda^2, \lambda^3)g(\lambda, \lambda^2, \lambda^3) = 0$ for all $\lambda \in \mathbf{A}^1$. Consider the one-variable polynomials $f_1(t) = f(t, t^2, t^3)$, $g_1(t) = g(t, t^2, t^3) \in \mathbb{R}[t]$, and let $V_{f_1} = \{\lambda \in \mathbf{A}^1 \mid f_1(\lambda) = f(\lambda, \lambda^2, \lambda^3) = 0\}$. If $V_{f_1} = \mathbf{A}^1$ then f vanishes on V and hence $f \in \mathbf{I}(V)$. If $V_{f_1} \neq \mathbf{A}^1$, then g_1 vanishes on the nonempty open subset $U = \mathbf{A}^1 - V_{f_1}$ of $\mathbf{A}^1$. It follows from Proposition 2.3(i) that g_1 vanishes on $\mathbf{A}^1$, and consequently g vanishes on V. This shows that $g \in \mathbf{I}(V)$, as desired. (Without using the density of U, the proof may also be completed by applying the infiniteness of U to a one-variable polynomial.)

Observe that the last example has used two important facts:

(a) There is a one-to-one and onto correspondence

$$\mathbf{A}^1 = K \longrightarrow V$$

$$\lambda \mapsto (\lambda, \lambda^2, \lambda^3)$$

(b) $\mathbf{A}^1$ is irreducible (so that Proposition 2.3(i) can be used).

The above observation shows the correct way to introduce polynomial mappings and rational mappings between algebraic sets, in order to compare the properties of two algebraic sets (such as irreducibility) with respect to the Zariski topology.

2.6. Definition Let $V \subseteq \mathbf{A}^m$, $W \subseteq \mathbf{A}^n$ be affine algebraic sets.
(i) If there exist polynomials $f_1, ..., f_n \in K[x_1, ..., x_m]$ that define a mapping

$$\phi: V \longrightarrow W$$

$$P \mapsto (f_1(P), ..., f_n(P))$$

then ϕ is called a *polynomial mapping* from V to W; if there is also a polynomial mapping $\psi: W \to V$ such that the composite mapping $\psi\phi = 1_V$, the identity mapping on V, and $\phi\psi = 1_W$, then ϕ is called a (polynomial) *isomorphism* with the inverse ψ (this name is qualified by Proposition 2.7 below).

If furthermore $V = \mathbf{A}^m$ then ϕ is called a *polynomial parametrization* of W.

(ii) If there exist elements $\frac{f_1}{g_1}, ..., \frac{f_n}{g_n} \in K(x_1, ..., x_m)$, where $K(x_1, ..., x_m)$ is the field of fractions of the polynomial ring $K[x_1, ..., x_n]$, that define a mapping

$$\phi: U = V - \bigcup_{i=1}^{n} \mathbf{V}(g_i) \longrightarrow W$$

$$P \mapsto \left(\frac{f_1(P)}{g_1(P)}, ..., \frac{f_n(P)}{g_n(P)} \right)$$

then ϕ is called a *rational mapping* from V to W; if furthermore $V = \mathbf{A}^m$ then ϕ is called a *rational parametrization* of W.

By definition, in Example (vi) above the correspondence $\lambda \mapsto (\lambda, \lambda^2, \lambda^3)$ defines a polynomial parametrization of the twisted cubic $V = \mathbf{V}(y - x^2, z - x^3)$.

2.7. Proposition (i) Polynomial and rational mappings are continuous with respect to Zariski topology.
(ii) Let $V \subseteq \mathbf{A}^m$, $W \subseteq \mathbf{A}^n$ be affine algebraic sets, and let ϕ be a (polynomial) rational mapping from V to W.
(a) Suppose ϕ is onto. If X is a closed subset of W and its preimage $\phi^{-1}(X)$ in V is irreducible, then X is irreducible.
(b) If $V = \mathbf{A}^m$ and the image of ϕ in W is dense, then W is irreducible.

Proof (i) It is sufficient to prove the assertion for rational mappings. Suppose ϕ is defined by

$$\phi: U = V - \bigcup_{i=1}^{n} \mathbf{V}(g_i) \longrightarrow W$$

$$P \mapsto \left(\frac{f_1(P)}{g_1(P)}, ..., \frac{f_n(P)}{g_n(P)} \right)$$

where $\frac{f_1}{g_1}, ..., \frac{f_n}{g_n} \in K(x_1, ..., x_m)$. If $W_1 = \mathbf{V}(h_1, ..., h_s) \cap W$ is a closed subset of W, where $h_1, ..., h_s \in K[y_1, ..., y_n]$, then $P \in U$ and $\phi(P) \in W_1$

implies for $i = 1, ..., s$

(1) $\quad 0 = h_i(\phi(P)) = h_i\left(\dfrac{f_1(P)}{g_1(P)}, ..., \dfrac{f_n(P)}{g_n(P)}\right) = h_i\left(\dfrac{f_1}{g_1}, ..., \dfrac{f_n}{g_n}\right)(P)$

Write $h_i\left(\dfrac{f_1}{g_1}, ..., \dfrac{f_n}{g_n}\right) = \dfrac{F_i}{G_i}$, $i = 1, ..., s$, where $F_i, G_i \in K[x_1, ..., x_m]$. Then it follows from equation (1) that $F_i(P) = 0$, $G_i(P) \neq 0$, $i = 1, ..., s$, and hence

(2) $\quad\quad\quad\quad\quad\quad\quad P \in \mathbf{V}(F_1, ..., F_s) \cap U.$

On the other hand, if $Q \in \mathbf{V}(F_1, ..., F_s) \cap U$, then $F_i(Q) = 0$ implies $\dfrac{F_i(Q)}{G_i(Q)} = 0$, $i = 1, ..., s$, and hence equation (1) holds for Q, that is, $\phi(Q) \in \mathbf{V}(h_1, ..., h_s) \cap W = W_1$. This shows that $\phi(\mathbf{V}(F_1, ..., F_s) \cap U) = W_1$, and consequently ϕ is continuous.

(ii) The assertion of part (a) follows from part (i) directly.

We prove (b) by showing that $\mathbf{I}(W)$ is a prime ideal. Let ϕ be defined by

$$\phi: U \longrightarrow V$$

$$P \mapsto \left(\dfrac{f_1(P)}{g_1(P)}, ..., \dfrac{f_n(P)}{g_n(P)}\right)$$

where $U = \mathbf{A}^m - (\cup_{i=1}^n \mathbf{V}(g_i))$, $\dfrac{f_1}{g_1}, ..., \dfrac{f_n}{g_n} \in K(y_1, ..., y_m)$, and $\phi(U)$ is dense in W by the assumption. Suppose $fg \in \mathbf{I}(W)$ where $f, g \in K[x_1, ..., x_n]$. Then $0 = (fg)(Q) = f(Q)g(Q)$ for all $Q \in W$. But this implies $0 = f\left(\dfrac{f_1}{g_1}, ..., \dfrac{f_n}{g_n}\right)(P)g\left(\dfrac{f_1}{g_1}, ..., \dfrac{f_n}{g_n}\right)(P)$ for all $P \in U$. Since $f\left(\dfrac{f_1}{g_1}, ..., \dfrac{f_n}{g_n}\right), g\left(\dfrac{f_1}{g_1}, ..., \dfrac{f_n}{g_n}\right) \in K(y_1, ..., y_m)$, we can write $f(\dfrac{f_1}{g_1}, ..., \dfrac{f_n}{g_n}) = \dfrac{F}{G}$, $g(\dfrac{f_1}{g_1}, ..., \dfrac{f_n}{g_n}) = \dfrac{F'}{G'}$, where $F, F', G, G' \in K[y_1, ..., y_m]$ with $G(P) \neq 0$ and $G'(P) \neq 0$ for all $P \in U$. Thus, if we put

$$V_f = \left\{P \in U \;\middle|\; f\left(\tfrac{f_1}{g_1}, ..., \tfrac{f_n}{g_n}\right)(P) = 0\right\}$$

$$V_g = \left\{P \in U \;\middle|\; g\left(\tfrac{f_1}{g_1}, ..., \tfrac{f_n}{g_n}\right)(P) = 0\right\}$$

then $V_f = \mathbf{V}(F) \cap U$, $V_g = \mathbf{V}(F') \cap U$. Now, if $V_f = U$, then $f(\phi(U)) = 0$ and hence $f(W) = 0$ because $\phi(U)$ is dense in W. So $f \in \mathbf{I}(W)$. If $V_f \neq U$, then since $(fg)(Q) = 0$ for all $Q \in W$, V_g, and consequently $\mathbf{V}(F')$ must contain the nonempty open subset $U - V_f = U - \mathbf{V}(F)$. It follows from

Proposition 2.3(i) that $V_g = U$. Therefore, $g(\phi(U)) = 0$ and $g(W) = 0$ by the density of $\phi(U)$ in W. Consequently, $g \in \mathbf{I}(W)$ as desired. $\square$

Remark (i) By Definition 2.6(i) and Proposition 2.7, algebraic sets may be classified in terms of "polynomial isomorphism". In a general theory of algebraic geometry, rational mappings provide models of abstract rational morphisms that yield classification of (quasi-)varieties in terms of "birational isomorphism".
(ii) Practically, polynomial parametrization and rational parametrization are very useful in 3D-plotting of curves and surfaces, geometric modelling, and CAD (computer aided design).

Since $K[x_1, ..., x_n]$ is Noetherian, it follows from Proposition 1.4(iv) that

- $\mathbf{A}^n$ is a *Noetherian space* in the sense that every collection of algebraic sets in $\mathbf{A}^n$ has a minimal member with respect to $\subseteq$ (*check it!*).

This observation enables us to prove the following decomposition theorem for affine algebraic sets.

2.8. Theorem Let V be an algebraic set in $\mathbf{A}^n$. Then V can be expressed uniquely as a union of finitely many irreducible algebraic sets $V_1, ..., V_m$, that is, $V = V_1 \cup \cdots \cup V_m$, and $V_i \not\subset V_j$ for all $i \neq j$.

Proof Let

$$\Omega = \left\{ \text{algebraic sets } V \subseteq \mathbf{A}^n \;\middle|\; \begin{array}{l} V \text{ is not a union of finitely many} \\ \text{irreducible algebraic sets} \end{array} \right\}.$$

We claim that Ω is empty. If not, let V be a minimal member of Ω. Since $V \in \Omega$, V is not irreducible. Thus, $V = V_1 \cup V_2$ where V_i are proper closed subsets of V, and $V_i \notin \Omega$. Hence $V_i = V_{i1} \cup \cdots \cup V_{im_i}$ with V_{ij} irreducible. But then $V = \cup_{i,j} V_{ij}$, a contradiction. Therefore $\Omega = \emptyset$. So any algebraic set V may be written as $V = V_1 \cup \cdots \cup V_m$, where each V_i is irreducible. To get the second condition, simply throw away any V_i such that $V_i \subset V_j$ for $i \neq j$.

To show uniqueness, let $V = W_1 \cup \cdots \cup W_m$ be another such decomposition. Then $V_i = \cup_j (W_j \cap V_i)$, so $V_i \subset W_{j(i)}$ for some $j(i)$. Similarly, $W_{j(i)} \subset V_k$ for some k. But $V_i \subset V_k$ implies $i = k$, so $V_i = W_{j(i)}$. Likewise each W_j is equal to some $V_{i(j)}$. $\square$

The V_i appearing in the theorem are called the *irreducible components* of V; $V = V_1 \cup \cdots \cup V_m$ is the *minimal decomposition* (or sometimes, the *irredundant union*) of V into irreducible components.

Example (vii) Consider the algebraic set $V = \mathbf{V}(xz, yz)$. Then V is a union of a line (the z-axis) and a plane (the xy-plane), both of which are irreducible.

(viii) Let $f \in K[x_1, ..., x_n]$ and $f = f_1^{n_1} f_2^{n_2} \cdots f_r^{n_r}$ the factorization of f into irreducible polynomials. If K is algebraically closed, then, by Nullstellensatz, it may be derived that $\mathbf{V}(f) = \mathbf{V}(f_1) \cup \cdots \cup \mathbf{V}(f_r)$ is the decomposition of $\mathbf{V}(f)$ into irreducible components. Moreover, it is an exercise to check that $\mathbf{I}(\mathbf{V}(f)) = \langle f_1 f_2 \cdots f_r \rangle$.

(ix) Note that any $P \in V = \mathbf{V}(xz - y^2, z^3 - x^5) \subset \mathbf{A}_\mathbb{R}^3$ is either the form $(\lambda^3, \lambda^4, \lambda^5)$ or the form $(-\lambda^3, -\lambda^4, -\lambda^5)$, where $\lambda \in \mathbb{R}$. If we plot a part of V, then it seems that V is a union of two irreducible curves.

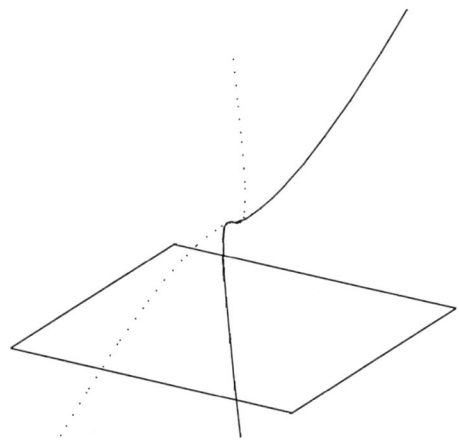

In later exercise 8 the reader will be asked to prove that it is the case indeed.

Theorem 2.8 can also be stated in a purely algebraic way using the one-to-one correspondence between radical ideals and algebraic sets.

2.9. Theorem If K is algebraically closed, then every radical ideal I in $K[x_1, ..., x_n]$ may be written uniquely as a finite intersection of prime ideals: $I = P_1 \cap \cdots \cap P_r$ where $P_i \not\subset P_j$ for $i \neq j$. (Such a presentation is called the *minimal decomposition* or the *irredundant intersection* of a radical ideal.)

Proof This follows immediately from Theorems 1.7 and 2.8. □

Remark Theorem 2.9 is indeed the geometric model of a general primary decomposition theory in commutative algebra.

We complete this section by seeking irreducible hypersurfaces $\mathbf{V}(f)$ defined by $f \in K[x_1, ..., x_n]$. If f is irreducible then the ideal $\langle f \rangle$ is a prime ideal, and hence $\sqrt{\langle f \rangle} = \langle f \rangle$.
(a) If K is algebraically closed, then $\mathbf{I}(\mathbf{V}(f)) = \sqrt{\langle f \rangle} = \langle f \rangle$ and hence $\mathbf{V}(f)$ is irreducible.
(b) If K is not algebraically closed, then exercise 2 of the current section shows that $\mathbf{V}(f)$ is not necessarily irreducible.
(c) If K is not algebraically closed but $f \in K[x, y]$ and $\mathbf{V}(f)$ is infinite, then $\mathbf{I}(\mathbf{V}(f)) = \langle f \rangle$ and consequently $\mathbf{V}(f)$ is irreducible by Theorem 2.4.

The assertion (c) is argued in detail as follows.

2.10. Proposition Let f and g be polynomials in $K[x, y]$ with no common divisors, and let $V = \mathbf{V}(f, g)$ be the algebraic set defined by f and g. Then $V = \mathbf{V}(f) \cap \mathbf{V}(g)$ is a finite set of points.

Proof f and g have no common divisors in $K[x][y]$, so they also have no common divisors in $K(x)[y]$ by Chapter 1 (section 2, exercise 7). Since $K(x)[y]$ is a PID, $pf + qg = 1$ for some $p, q \in K(x)[y]$. Let $d \in K[x]$ be such that $dp = a$, $dq = b \in K[x, y]$. Then $af + bg = d$. If $P = (u, v) \in V$, then $d(P) = d(u) = 0$. But d has only a finite number of zeros. This shows that only a finite number of x-coordinates appear among the points of V. Since the same reasoning applies to the y-coordinates, there can only be a finite number of points. □

2.11. Corollary If f is an irreducible polynomial in $K[x, y]$, and if $\mathbf{V}(f)$ is infinite, then $\mathbf{I}(\mathbf{V}(f)) = \langle f \rangle$ and $\mathbf{V}(f)$ is irreducible.

Proof If $g \in \mathbf{I}(\mathbf{V}(f))$ then $\mathbf{V}(f, g)$ is infinite. By the proposition, $f | g$, i.e., $g \in \langle f \rangle$. □

2.12. Corollary Suppose K is infinite. The irreducible algebraic subsets of $\mathbf{A}^2$ are: $\mathbf{A}^2$, points, and irreducible plane curves $\mathbf{V}(f)$, where f is an irreducible polynomial and $\mathbf{V}(f)$ is infinite. (Note that we do not view $\emptyset$ as an irreducible algebraic set, as we indicated in Definition 2.2.)

Proof Let V be an irreducible algebraic set in $\mathbf{A}^2$. If V is finite or $\mathbf{I}(V) = (0)$, V is of the required type. Otherwise, $\mathbf{I}(V)$ contains an irreducible polynomial f because $\mathbf{I}(V)$ is now a prime ideal. Then we claim $\mathbf{I}(V) = \langle f \rangle$; for if $g \in \mathbf{I}(V)$, $f \nmid g$, then $V \subset \mathbf{V}(f, g)$ is finite by Proposition 2.10, contradicting the infiniteness of V. □

Appendix. Topological space

Given a nonempty set X, a *topology* on X is a collection T of subsets of X satisfying
(a) $\emptyset \in T$, $X \in T$,
(b) the union of any family of sets in T belongs to T, and
(c) the intersection of finitely many sets in T is also in T.
The sets in T are called *open sets* of X, and their complements in X are called *closed sets*.

A set X equipped with a topology T is then called a *topological space*, and the elements of X are usually called *points*.

Given a topological space X, a *neighborhood* of a point $x \in X$ is any set that contains some open set containing x.

Example (i) Any set X can be made into a topological space by giving the discrete topology, in which every subset (including empty set) is open.

(ii) The real line $\mathbb{R}$ is a topological space, where a set U is open if it is a union of (finitely or infinitely many) open intervals.

(iii) The real Euclidean space $\mathbb{R}^n$ is a topological space, where the open sets are defined as the union of open balls.

(iv) Any metric space X is equipped with the topology in which open sets are unions of open balls.

(v) Every manifold (if you know what it is) is a topological space.

Topological spaces appear in all branches of modern mathematics, in which it is necessary to study the convergence, connectedness and continuity.

Let X and Y be topological spaces and $f\colon X \to Y$ a mapping. f is said to be a *continuous function* if the preimage of every open set in Y is an open set in X. Suppose that f is a one-to-one and onto mapping with inverse f^{-1}. If both f and f^{-1} are continuous, then f is called a *homeomorphism*, and in this case, X and Y are said to be *homeomorphic*. Homeomorphic topological spaces have essentially the same topological structure.

Any subset Y of a topological space X has the *induced topology*, in which, every open set is of the form $Y \cap U$ where U is an open set of X.

Given topological spaces $X_1, X_2, ..., X_n$, the Cartesian product $X_1 \times X_2 \times \cdots \times X_n$ has the *product topology* given by the products of open sets.

Finally, let X be a topological space, Y a subset of X. The *closure* of Y in X, denoted $\overline{Y}$, is the smallest closed subset in X containing Y, i.e., $\overline{Y} = \cap W$ where W runs over all closed subsets of X containing Y. A subset Z of X is said to be *dense* in X if $Z \cap U \neq \emptyset$ for every nonempty open subset $U \subset X$, or equivalently, if $\overline{Z} = X$.

Exercises

1. Show that $V = \mathbf{V}(y - f(x)) \subset \mathbf{A}_\mathbb{C}^2$ is irreducible.
2. (This exercise and the next help us to better understand the Nullstellensatz.) Show that $f = y^2 + x^2(x-1)^2 \in \mathbb{R}[x, y]$ is an irreducible polynomial, but that $\mathbf{V}(f)$ is reducible.
3. Show that if $f \in \mathbb{C}[x_1, ..., x_n]$ is irreducible, then $V(f)$ is irreducible. Also show that if $V = \mathbf{V}(g)$ is an irreducible hypersurface in $\mathbf{A}_\mathbb{C}^n$, there is no irreducible algebraic set W such that $V \subset W \subset \mathbf{A}_\mathbb{C}^n$, $W \neq V$, $W \neq \mathbf{A}_\mathbb{C}^n$.
4. Show that any linear subspace of $K^n = \mathbf{A}_K^n$, where K is a field, is an irreducible algebraic set. (Hint: Use Proposition 2.7.)
5. Use exercise 3 above to show that $\mathbf{V}(y^2 - x(x-1)(x-\lambda))$ is an irreducible curve in $\mathbf{A}_\mathbb{C}^2$, where $\lambda \in \mathbb{C}$.
6. Let $\phi\colon \mathbf{A}_\mathbb{R}^1 \to V = \mathbf{V}(y^2 - x^3 - x^2)$ be defined by $\phi(c) = (c^2 - 1, c(c^2 - 1))$. Show that ϕ is one-to-one and onto, except that $\phi(\pm 1) = (0, 0)$. Hence V is irreducible by Proposition 2.7 (see also section 6, Example (i)).
7. Find the irreducible components of $\mathbf{V}(y^2 - xy - x^2 y + x^3)$ in $A_\mathbb{R}^2$, and also in $A_\mathbb{C}^2$.
8. Let $I = \langle xz - y^2, z^3 - x^5 \rangle$. Show that $\mathbf{V}(I)$ is a union of two irreducible algebraic sets.
9. Let $f \in \mathbb{C}[x_1, ..., x_n]$ and let $f = f_1^{n_1} \cdots f_r^{n_r}$ be the decomposition of f into irreducible factors. Show that $\mathbf{V}(f) = \mathbf{V}(f_1) \cup \cdots \cup \mathbf{V}(f_r)$ is

the decomposition of $\mathbf{V}(f)$ into irreducible components and $\mathbf{I}(\mathbf{V}(f)) = \langle f_1 f_2 \cdots f_r \rangle$.
10. Give a detailed proof of Proposition 2.3(ii) and Proposition 2.5, respectively.
11. Let $V \subseteq \mathbf{A}^m$, $W \subseteq \mathbf{A}^n$ be algebraic sets. If $\phi\colon V \to W$ is a polynomial mapping defined by $P \mapsto (f_1(P), ..., f_n(P))$ where $f_1, ..., f_n \in K[x_1, ..., x_m]$, show that $\phi_*\colon K[W] \to K[V]$ defined by $\phi_*(\overline{g}) = \overline{g(f_1, ... f_n)}$ is a K-linear ring homomorphism (i.e., a K-algebra homomorphism); if furthermore ϕ is an isomorphism then ϕ_* is a ring isomorphism.
Conversely, if $\psi\colon K[W] \to K[V]$ is a K-linear ring homomorphism such that $\psi(\overline{y}_i) = \overline{f}_i$ $i = 1, ..., n$, show that $\psi^*\colon V \to W$ defined by $\psi^*(P) = (f_1(P), ..., f_n(P))$ is a polynomial mapping.
Thus we may conclude that ϕ is an isomorphism if and only if ϕ_* is an isomorphism.

3. Point P and the Local Ring $\mathcal{O}_{P,V}$

By Theorem 2.8, the study of an algebraic set is reduced to the study of its irreducible components. In this section we demonstrate how a point P of an irreducible algebraic set V determines a unique local ring $\mathcal{O}_{P,V}$ consisting of rational functions defined at P.

Let $V \subseteq \mathbf{A}^n = \mathbf{A}_K^n$ be an *irreducible* algebraic set, $\mathbf{I}(V)$ its ideal, and $K[V] = K[x_1, ..., x_n]/\mathbf{I}(V)$ its coordinate ring. For $f, g \in K[x_1, ..., x_n]$, write $\overline{f}, \overline{g}$ for their images in $K[V]$. Then $\overline{f} = \overline{g}$ if and only if $f - g \in \mathbf{I}(V)$. Thus, every $\overline{f} \in K[V]$ defines a function

$$\phi_{\overline{f}}\colon V \longrightarrow K$$
$$P \mapsto f(P)$$

which is independent of the choice of representatives of $\overline{f}$. Moreover, let W be a closed subset of $K = \mathbf{A}^1$ with respect to the Zariski topology on $\mathbf{A}^1$, i.e., $W = \mathbf{V}(g)$ for some $g \in K[x]$ and W is finite. Then one checks that

$$\phi_{\overline{f}}(\mathbf{V}(g(f)) \cap V) = W.$$

This shows that $\phi_{\overline{f}}$ is a continuous function with respect to the Zariski topology. $\phi_{\overline{f}}$ is called a *polynomial function* on V. Note that if $P = (a_1, ..., a_n) \in V$, then $\phi_{\overline{x}_i}(P) = a_i$ for $i = 1, ..., n$. And since $K[V]$ is generated by all $\bar{x}_i$'s over K, this qualifies the name "coordinate ring" $K[V]$.

Also note that $K[V]$ is a domain by Theorem 2.4. We seek more continuous functions on V by looking at the field of fractions of $K[V]$, denoted $K(V)$. For $0 \neq \frac{\bar{f}}{g} \in K(V)$, set

$$J_{\frac{\bar{f}}{g}} = \left\{ d \in K[x_1, ..., x_n] \ \middle| \ \bar{d} \cdot \frac{\bar{f}}{g} \in K[V] \right\}.$$

Then $J_{\frac{\bar{f}}{g}}$ is an ideal of $K[x_1, ..., x_n]$ and $\mathbf{I}(V) \subseteq J_{\frac{\bar{f}}{g}}$ (check this!).

3.1. Lemma $\mathbf{V}\left(J_{\frac{\bar{f}}{g}}\right) = \left\{ P \in V \ \middle| \ \frac{\bar{f}}{g} = \frac{\bar{h}}{d} \text{ implies } d(P) = 0 \right\}.$

Proof By definition, this can be verified directly. □

The algebraic set $\mathbf{V}\left(J_{\frac{\bar{f}}{g}}\right)$ is called the *pole set* of $\frac{\bar{f}}{g}$ in V.

Example (i) Let $V = \mathbf{V}(y^3 - xz) \subset \mathbf{A}^3$. Then for $\frac{\bar{f}}{g} = \frac{\bar{x}}{y} = \frac{\bar{y}^2}{z} \in K(V)$,

$\mathbf{V}\left(J_{\frac{\bar{f}}{g}}\right) = \{(x, y, z) \in \mathbf{A}^3 \mid y = z = 0\}.$

For $\frac{\bar{f}}{g} \in K(V)$, let

$$U_{\frac{\bar{f}}{g}} = V - \mathbf{V}\left(J_{\frac{\bar{f}}{g}}\right).$$

Note that if $V \neq \emptyset$ then $U_{\frac{\bar{f}}{g}} \neq \emptyset$ (why?). Thus, $\frac{\bar{f}}{g}$ defines a function

$$\phi_{\frac{\bar{f}}{g}} : U_{\frac{\bar{f}}{g}} \longrightarrow K$$

$$P \mapsto \frac{f(P)}{g(P)}$$

which is independent of the choice of representatives of $\frac{\bar{f}}{g}$. Moreover, if $W = \mathbf{V}(h)$ for some $\sum \lambda_i x^i = h \in K[x]$, and thus $h(f/g) = f_1/g^m$, then one checks that

$$\phi_{\frac{\bar{f}}{g}}(\mathbf{V}(f_1) \cap U) = W,$$

i.e., $\phi_{\frac{f}{g}}$ is continuous on $U_{\frac{f}{g}}$.

By Proposition 2.3(i), the open subset $U_{\frac{f}{g}}$ is dense in V. Therefore, from a topological point of view, $\phi_{\frac{f}{g}}$ is "locally" defined "almost everywhere" on V. It follows that $K(V)$ is also called the *field of rational functions* of V. The *key* property of $\phi_{\frac{f}{g}}$ is that it is an "analytic-like" function in the sense of the next proposition.

3.2. Proposition Let V and $k(V)$ be as before. Two rational functions $\phi_{\frac{f}{g}}$ and $\phi_{\frac{h}{d}}$ of V are equal if and only if they agree on some open subset $U \subseteq V$.

Proof Suppose $\phi_{\frac{f}{g}}$ and $\phi_{\frac{h}{d}}$ agree on U. Then $U \subseteq U_{\frac{f}{g}} \cap U_{\frac{h}{d}}$, and $\phi_{\frac{f}{g}}(P) = \phi_{\frac{h}{d}}(P)$ implies $(gh - df)(P) = 0$ for all $P \in U$. Let $s = gh - df$ and $\psi_s : V \to K$ be the polynomial function defined by s. Then $U \subseteq W$, where W is the preimage of $\{0\}$ under ψ_s. Note that $\{0\}$ is closed in $K = \mathbf{A}^1$ and U is dense in V. It follows that $W = V$ and $s \in \mathbf{I}(V)$. Thus, in $K[V]$, $0 = \overline{s} = \overline{gh} - \overline{df}$, and hence $\phi_{\frac{f}{g}} = \phi_{\frac{h}{d}}$. □

The nice "analytic-like" property of a rational function discussed above makes the basis to define abstract regular functions on a quasi-variety in modern algebraic geometry in the schematic language.

Now, we can associate to each $P \in V$ a subring of $K(V)$ as follows. Notation is maintained as before.

3.3. Definition For $P \in V$, we say that $\phi_{\frac{f}{g}}$ is *defined at* P if $P \in U_{\frac{f}{g}}$.

Remark In fact, this definition says that $\phi_{\frac{f}{g}}$ is defined at P if there are some $h, s \in K[x_1, ..., x_n]$ such that $s(P) \neq 0$ and $\frac{\overline{h}}{\overline{s}} = \frac{\overline{f}}{\overline{g}}$. For instance, consider $P = (0, 0) \in V = \mathbf{V}(y^2 - x^3) \subset \mathbf{A}^2_K$, $\left(\frac{\overline{y}}{\overline{x}}\right)^2 = \overline{x}$ in $K(V)$.

Put

$$\mathcal{O}_{P,V} = \left\{ \frac{\overline{f}}{\overline{g}} \in k(V) \;\middle|\; \phi_{\frac{f}{g}} \text{ is defined at } P \right\}.$$

Then it is clear that $k \subset k[V] \subset \mathcal{O}_{P,V} \subset k(V)$.

3.4. Theorem (i) $\mathcal{O}_{P,V}$ is a local ring with the maximal ideal

$$M_{P,V} = \left\{ \frac{\overline{f}}{\overline{g}} \in \mathcal{O}_{P,V} \;\middle|\; \phi_{\frac{\overline{f}}{\overline{g}}}(P) = 0 \right\}.$$

Moreover, $\mathcal{O}_{P,V}$ is uniquely determined by P.
(ii) (Compare with Chapter 2 Theorem 3.8.) If K is algebraically closed, then

$$K[V] = \bigcap_{P \in V} \mathcal{O}_{P,V},$$

i.e., the rational functions that are defined at every $P \in V$ are nothing but all polynomial functions on V.

Proof (i) That $\mathcal{O}_{P,V}$ is a subring of $K(V)$ may be verified directly. Note that if $\phi_{\frac{\overline{f}}{\overline{g}}}(P) \neq 0$ then $\frac{\overline{f}}{\overline{g}}$ is invertible in $\mathcal{O}_{P,V}$. Hence $\mathcal{O}_{P,V}$ is a local ring with the described maximal ideal $M_{P,V}$. Indeed, this result can also be obtained by using the surjective ring homomorphism

$$\mathcal{O}_{P,V} \longrightarrow K$$

$$\frac{\overline{f}}{\overline{g}} \mapsto \phi_{\frac{\overline{f}}{\overline{g}}}(P)$$

For $P = (a_1, ..., a_n) \in V$, $x_i - a_i \in M_{P,V}$, $i = 1, ..., n$. If $Q = (b_1, ..., b_n) \in V$ and $Q \neq P$, then it is clear that $M_{P,V} \neq M_{Q,V}$ and hence $\mathcal{O}_{P,V} \neq \mathcal{O}_{Q,V}$. This proves (i).
(ii) If $\frac{\overline{f}}{\overline{g}} \in \cap_{P \in V} \mathcal{O}_{P,V}$, then $\mathbf{V}(J_{\frac{\overline{f}}{\overline{g}}}) = \emptyset$, so $1 \in J_{\frac{\overline{f}}{\overline{g}}}$ by Nullstellensatz, i.e., $1 \cdot \frac{\overline{f}}{\overline{g}} = \frac{\overline{f}}{\overline{g}} \in k[V]$, as desired. $\square$

3.5. Definition $\mathcal{O}_{P,V}$ is called the *local ring* of V at P.

3.6. Proposition $\mathcal{O}_{P,V} = K[V]_M$, where the latter is the localization of $K[V]$ at the maximal ideal

$$M = \left\{ \overline{f} \in K[V] \;\middle|\; \phi_{\overline{f}}(P) = 0 \right\}.$$

Hence $\mathcal{O}_{P,V}$ is Noetherian.

Proof If $P = (a_1, ..., a_n) \in V$, then

$$M = \langle x_1 - a_1, ..., x_n - a_n \rangle (\mathrm{mod}\ \mathbf{I}(V))$$

by Chapter 2 (section 1, Example (iii)). In view of Chapter 2 section 3, $K[V]_M$ is a subring of $K(V)$ and it is (by definition) nothing but $\mathcal{O}_{P,V}$. Thus, $\mathcal{O}_{P,V}$ is Noetherian by Corollary 3.5(ii) of Chapter 2. $\square$

The last proposition qualifies Definition 3.7 of Chapter 2.

Exercises

1. Verify Lemma 3.1.
2. Let $V = \mathbf{V}(y^2 - x^3 - x^2) \subset \mathbf{A}^2_{rz}$ be the nodal curve, and $\bar{x}, \bar{y}$ the image of x, y in $k[V]$; let $z = \frac{\bar{y}}{\bar{x}} \in k(V)$. Find the pole sets of z and of z^2. (Note that V is irreducible by (section 2, Exercise 6).)
3. Let $\mathcal{O}_{P,V}$ be the local ring of an irreducible algebraic set V at a point P. Show that there is a natural one-to-one correspondence between the prime ideals of $\mathcal{O}_{P,V}$ and the irreducible algebraic subsets of V containing P. (Hint: If I is prime in $\mathcal{O}_{P,V}$, $I \cap k[V]$ is prime in $k[V]$, and I is generated by $I \cap k[V]$; use Theorem 3.4.)
4. Let V be an irreducible algebraic set in $\mathbf{A}^n$, $I = \mathbf{I}(V)$, $P \in V$, and let J be an ideal of $K[x_1, ..., x_n]$ which contains I. Let $\overline{J}$ be the image of J in $k[V]$. Show that there is a natural homomorphism φ from $\mathcal{O}_{P,\mathbf{A}^n}/J\mathcal{O}_{P,\mathbf{A}^n}$ to $\mathcal{O}_{P,V}/\overline{J}\mathcal{O}_{P,V}$, and φ is an isomorphism. In particular, $\mathcal{O}_{P,\mathbf{A}^n}/I\mathcal{O}_{P\mathbf{A}^n}$ is isomorphic to $\mathcal{O}_{P,V}$.
   Also show that if $V = \mathbf{V}(I) = \{P\}$, then

   $$K[x_1, ..., x_n]/I \cong \mathcal{O}_{P,\mathbf{A}^n}/I\mathcal{O}_{P,\mathbf{A}^n}.$$

   (Hint: Use Chapter 2 Corollary 4.4.)
5. Let $V \subseteq \mathbf{A}^m$, $W \subseteq \mathbf{A}^n$ be irreducible algebraic sets. If $\phi: U \to W$ is a rational mapping defined by $P \mapsto (\alpha_1(P), ..., \alpha_n(P))$ where $\alpha_1 = \frac{f_1}{g_1}, ..., \alpha_n = \frac{f_n}{g_n} \in K(x_1, ..., x_m)$ and $U = V - \cup_{i=1}^n \mathbf{V}(g_i)$, show that $\phi_*: K(W) \to K(V)$ defined by $\phi_*\left(\frac{F}{G}\right) = \overline{F(\alpha_1, ..., \alpha_n)/G(\alpha_1, ..., \alpha_n)}$ is a K-linear ring homomorphism.
   Conversely, if $\psi: K(W) \to K(V)$ is a K-linear ring homomorphism such that $\psi(\bar{x}_i) = \alpha_i = \frac{f_i}{g_i} \in K(x_1, ..., x_m)$, $i = 1, ..., n$, then show that $P \mapsto (\alpha_1(P), ..., \alpha_n(P))$ defines a rational mapping $\psi^*: U \to W$ where $U = V - \cup_{i=1}^n \mathbf{V}(g_i)$.

4. Nonsingular Points and DVRs

This section is devoted to showing how to algebraically recognize singularities in algebraic curves.

Let $\mathbf{A}^2 = \mathbf{A}_K^2$. If $f \in K[x,y]$ is a *nonconstant* polynomial, then we write $\mathcal{C} = \mathbf{V}(f)$ for the plane curve defined by f (and hence by cf for any $0 \neq c \in K$), and $K[\mathcal{C}] = K[x,y]/\mathbf{I}(\mathcal{C})$ for its coordinate ring; in the case where $\mathcal{C}$ is irreducible we also write $K(\mathcal{C})$ for its field of rational functions. If f has degree $n \geq 1$, we say that the curve $\mathcal{C}$ is of *degree n*.

To have a more convenient discussion, the first thing to be explored is that after an (affine) change of coordinates nothing about $\mathcal{C} = \mathbf{V}(f)$ is "intrinsically" changed. (Although there is a result that applies polynomial isomorphisms to general affine algebraic sets, we do not have space to discuss it.)

Let $a, b \in K$. Define the (affine) change of coordinates (as in $\mathbb{R}^2$)

$$\phi: \quad \mathbf{A}^2 \longrightarrow \mathbf{A}^2$$
$$(x,y) \mapsto (x+a, y+b)$$

and put $\mathcal{C}' = \mathbf{V}(f(x+a, y+b))$.

4.1. Lemma With notation as above, $\phi(\mathcal{C}') = \mathcal{C}$.

Proof Exercise. $\square$

4.2. Proposition (i) The foregoing ϕ induces a K-linear ring isomorphism $\phi_*: K[\mathcal{C}] \longrightarrow K[\mathcal{C}']$.
(ii) If $\mathcal{C}$ is irreducible, then ϕ induces a K-linear ring isomorphism $\overline{\phi}_*: K(\mathcal{C}) \longrightarrow K(\mathcal{C}')$ that yields a ring isomorphism $\mathcal{O}_{\phi(Q),\mathcal{C}} \longrightarrow \mathcal{O}_{Q,\mathcal{C}'}$ with $\overline{\phi}_*(M_{\phi(Q),\mathcal{C}}) = M_{Q,\mathcal{C}'}$ for every $Q \in \mathcal{C}'$.

Proof (i) Consider the K-linear ring homomorphism

$$\phi_*: \quad K[x,y] \longrightarrow K[x,y]$$
$$g(x,y) \mapsto g(x+a, y+b)$$

It is well-known that ϕ_* is an isomorphism (or use (section 2, exercise 11) as ϕ is now an isomorphism). Let $h \in I = \mathbf{I}(\mathcal{C})$. Then $\phi_*(h) = h(x+a, y+b)$. For any $Q = (c,d) \in \mathcal{C}'$, $0 = h(\phi(Q)) = h(c+a, d+b) = \phi_*(h)(Q)$ by Lemma 4.1. This shows that $\phi_*(I) \subseteq I' = \mathbf{I}(\mathcal{C}')$. Conversely, if $g \in I'$, then since $(u,v) \in \mathcal{C}$ implies $(u-a, v-b) \in \mathcal{C}'$, we have $g(u-a, v-b) = 0$. It follows that $g_1 = g(x-a, y-b) \in I$ and $\phi_*(g_1) = g \in I'$. Therefore, $\phi_*(I) = I'$ and

$$K[\mathcal{C}] = \frac{K[x,y]}{I} \xrightarrow[\cong]{\phi_*} \frac{K[x,y]}{I'} = K[\mathcal{C}']$$

(ii) If $\mathcal{C}$ is irreducible, by part (i) we have a K-linear ring isomorphism induced by ϕ_*

$$\overline{\phi}_* : K(\mathcal{C}) \longrightarrow K(\mathcal{C}')$$

$$\frac{f}{g} \mapsto \frac{\phi_*(f)}{\phi_*(g)}$$

It is not hard to check that $\mathcal{O}_{\phi(Q),\mathcal{C}} \xrightarrow[\cong]{\overline{\phi}_*} \mathcal{O}_{Q,\mathcal{C}'}$ with $\overline{\phi}_*(M_{\phi(Q),\mathcal{C}}) = M_{Q,\mathcal{C}'}$ for every $Q \in \mathcal{C}'$. □

For a curve $\mathcal{C} = \mathbf{V}(f) \subset \mathbf{A}^2$, in order to introduce the singularity in $\mathcal{C}$, we first provide an algebraic way to define the tangent line of $\mathcal{C}$ at a point P.

Let L be a line passing $P = (a,b) \in \mathcal{C}$, which is parametrized as

$$\begin{cases} x = a + ct \\ y = b + dt \end{cases}$$

where t is a variable. Put $g(t) = f(a + ct, b + dt)$. Then $t = 0$ is a root of $g(t)$. Considering the formal MacLaurin series expansion of $g(t)$ at 0, we have

$$g(t) = g(0) + \frac{g'(0)}{1!}t + \cdots + \frac{g^{(n)}(0)}{n!}t^n.$$

If $g^{(m)}(0) \neq 0$, and $g^{(\ell)}(0) = 0$ for $\ell < m$, then 0 is a root of multiplicity m for $g(t)$, and we say that L intersects $\mathcal{C}$ at $P = (a,b)$ *with multiplicity m*. It is easy to show that this definition is independent of the particular parametrization of the line L.

4.3. Proposition Put $\nabla f = \left(\dfrac{\partial f}{\partial x}, \dfrac{\partial f}{\partial y}\right)$ (the gradient vector of f).
(i) If $\nabla f(P) \neq (0,0)$, then there is a *unique* line through P which intersects $\mathcal{C}$ with multiplicity ≥ 2.

(ii) If $\nabla f(P) = (0,0)$, then *every* line through P intersects $\mathcal{C}$ with multiplicity ≥ 2.

Proof Note that $t = 0$ is a root of multiplicity ≥ 2 if and only if $g'(0) = 0$ if and only if

(1) $$\frac{\partial f}{\partial x}(P) \cdot c + \frac{\partial f}{\partial y}(P) \cdot d = 0.$$

Hence, if $\nabla f(P) = (0,0)$, then L always intersects $\mathcal{C}$ at P with multiplicity ≥ 2. This proves (ii).

Now suppose $\nabla f(P) \neq (0,0)$. Then the solution space of the equation (1) with unknowns c and d is 1-dimensional. Thus, there is $(c_0, d_0) \neq (0,0)$ with the property that (c,d) satisfies (1) if and only if $(c,d) = \lambda(c_0, d_0)$ for some $\lambda \in k$. It follows that the pairs (c,d) that make $g'(0) = 0$ all parametrize the same line. This shows that there is a unique line which intersects $\mathcal{C}$ at P with multiplicity ≥ 2, and hence (i) is proved. □

4.4. Definition If $\nabla f(P) \neq (0,0)$, then the *tangent line* of $\mathcal{C}$ at P is defined to be the unique line through P intersecting $\mathcal{C}$ with multiplicity ≥ 2 at P. In this case, P is called a *nonsingular point* (or a *simple point*) of $\mathcal{C}$.

If $\nabla f(P) = (0,0)$, then P is called a *singular point* of $\mathcal{C}$, or we say that P defines a *singularity* in $\mathcal{C}$.

If every point $P \in \mathcal{C}$ is nonsingular, then $\mathcal{C}$ is called a *nonsingular curve* (or a *smooth curve*).

Example (i) Any irreducible conic $\mathcal{C} \subset \mathbf{A}_{\mathbb{C}}^2$ is nonsingular. To see this, let $P \in \mathcal{C}$. After a change of coordinates we may assume $P = (0,0)$. If $\frac{\partial f}{\partial x}(P) = \frac{\partial f}{\partial y}(P) = 0$, then f would be of the form $ax^2 + by^2 + cxy$, which can be factorized into a product of linear forms.

(ii) $P = (0,0)$ is a singular point in the cuspidal curve $\mathcal{C} = \mathbf{V}(y^2 - x^3)$

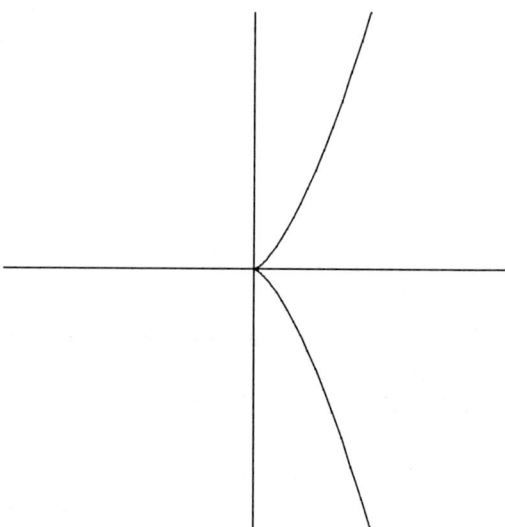

and $P = (0,0)$ is also a singular point in the nodal curve $\mathcal{C} = \mathbf{V}(y^2 - x^3 - x^2)$.

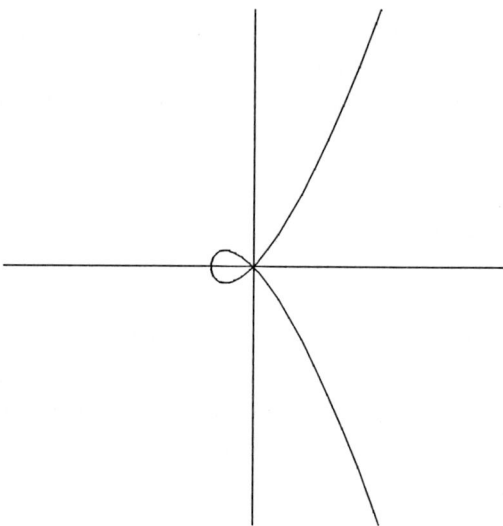

Let $\mathcal{C} = \mathbf{V}(f) \subset \mathbf{A}^2$ be a curve of degree $n \geq 1$, $P = (0,0) \in \mathcal{C}$. Then f

must be of the form

$$f = F_m + F_{m+1} + \cdots + F_q \text{ with } m \geq 1, \ F_j = \sum_{\alpha+\beta=j} \lambda_{\alpha\beta} x^\alpha y^\beta,$$

i.e., every F_j is a homogeneous polynomial of degree $j \geq m \geq 1$. In view of the previous discussion on singularity, this number m is usually called the *multiplicity* of $\mathcal{C}$ at $P = (0,0)$, denoted $m_P(\mathcal{C}) = m$.

Observation $P = (0,0)$ is a simple point if and only if $m = 1$.

If $\mathcal{C}$ is irreducible, then $K \cong \mathcal{O}_{P,\mathcal{C}}/M_{P,\mathcal{C}}$ by the proof of Theorem 3.4(i). So $M_{P,\mathcal{C}}^n/M_{P,\mathcal{C}}^{n+1}$ may be regarded as a K-vector space for every $n \geq 0$. This enables us to characterize $m_P(\mathcal{C})$ algebraically.

4.5. Theorem Let $\mathcal{C} = \mathbf{V}(f) \subset \mathbf{A}^2$ be an irreducible curve of degree $n \geq 1$, $P = (0,0) \in \mathcal{C}$. Write $\mathcal{O}$ for the local ring of $\mathcal{C}$ at P and M for the maximal ideal of $\mathcal{O}$. If $\mathcal{C}$ has the multiplicity $m_P(\mathcal{C}) = m$ as defined above, then

$$m_P(\mathcal{C}) = \dim_k \left(M^n/M^{n+1} \right) \quad \text{for large } n.$$

Proof From the isomorphism

$$\left((\mathcal{O}/M^{n+1}) \big/ (M^n/M^{n+1}) \right) \cong (\mathcal{O}/M^n)$$

it follows that it is sufficient to prove that

$$\dim_K (\mathcal{O}/M^n) = n \cdot m_P(\mathcal{C}) + s$$

for some constant s and all $n \geq m = m_P(\mathcal{C})$. Since $P = (0,0)$, we have $M^n = \mathbf{m}^n \mathcal{O}$, where $\mathbf{m} = \langle x, y \rangle \subset K[x,y]$. Note that $\mathbf{V}(\mathbf{m}^n) = \{P\} = \mathbf{V}(\mathbf{m}^n, f)$. By section 3, exercise 5,

$$\frac{K[x,y]}{\langle \mathbf{m}^n, f \rangle} \cong \frac{\mathcal{O}_{P,\mathbf{A}^2}}{\langle \mathbf{m}^n, f \rangle \mathcal{O}_{P,\mathbf{A}^2}} \cong \frac{\mathcal{O}}{\mathbf{m}^n \mathcal{O}} = \frac{\mathcal{O}}{M^n}$$

where $\mathcal{O}_{P,\mathbf{A}^2} \subset K(x,y)$ is the local ring of $\mathbf{A}^2$ at P. Thus, the problem is reduced to the calculation of the dimension of $K[x,y]/\langle \mathbf{m}^n, f \rangle$. Since $m = m_P(\mathcal{C})$ implies $fg \in \mathbf{m}^n$ whenever $g \in \mathbf{m}^{n-m}$, there is a natural ring homomorphism $\varphi: K[x,y]/\mathbf{m}^n \to K[x,y]/\langle \mathbf{m}^n, f \rangle$, and a K-linear mapping $\psi: K[x,y]/\mathbf{m}^{n-m} \to K[x,y]/\mathbf{m}^n$ defined by $\psi(\bar{g}) = \overline{fg}$. It is easy to verify that $\varphi \circ \psi$ induces the isomorphism of K-vector spaces:

$$\left(\left(\frac{K[x,y]}{\mathbf{m}^n}\right)\Big/\left(\frac{\langle f\rangle + \mathbf{m}^n}{\mathbf{m}^n}\right)\right) \cong \frac{K[x,y]}{\langle \mathbf{m}^n, f\rangle}.$$

It follows from section 1, Exercise 14 that

$$\dim_K \left(\frac{K[x,y]}{\langle \mathbf{m}^n, f\rangle}\right) = nm - \frac{m(m-1)}{2}, \quad \text{for all } n \geq m,$$

as desired. □

Remark (i) By the theorem, $m_P(\mathcal{C})$ depends only on the local ring of $P = (0,0)$, that is, it is an intrinsic property. Considering any point $Q = (a,b) \in \mathcal{C}$, by Lemma 4.1 and Proposition 4.2, the multiplicity (for singularity) of $\mathcal{C}$ at Q, denoted $m_Q(\mathcal{C})$, may be defined to be $m_{(0,0)}(\mathcal{C}')$, where $\mathcal{C}' = \mathbf{V}(f(x+a, y+b))$.

(ii) It can be shown that the function $\chi(n) = \dim_K \left(\mathcal{O}_{P,\mathcal{C}}/M_{P,\mathcal{C}}^n\right)$ is a polynomial in n (for large n) which is called the Hilbert-Samuel polynomial of the local ring $\mathcal{O}_{P,\mathcal{C}}$.

We are ready to reach the main result of this section.

4.6. Theorem Let $\mathcal{C} = \mathbf{V}(f) \subset \mathbf{A}^2$ be an irreducible curve of degree $n \geq 1$, $P = (a,b) \in \mathcal{C}$. Then P is a simple point if and only if the local ring $\mathcal{O}_{P,\mathcal{C}}$ of $\mathcal{C}$ at P is a DVR in $K(\mathcal{C})$, and in this case, the maximal ideal $M_{P,\mathcal{C}}$ of $\mathcal{O}_{P,\mathcal{C}}$ can be generated by $\overline{L}$ where $\mathbf{V}(L = ax + by + c)$ is any line through P but not tangent to $\mathcal{C}$ at P.

Proof First suppose that P is a simple point in $\mathcal{C}$. Since L is not the tangent line of $\mathcal{C}$ at P, after an appropriate (affine) change of coordinates we may assume that $P = (0,0)$, that y is the tangent line, and that $L = x$ by Proposition 4.2. Then by Chapter 2 Theorem 2.10, it is sufficient to show that $M_{P,\mathcal{C}}$ is generated by the image of x in $\mathcal{O}_{P,\mathcal{C}}$.

Write $\overline{x}, \overline{y}$ for the images of x, y in $k[\mathcal{C}]$. Then $P = (0,0)$ implies $M_{P,\mathcal{C}} = \langle \overline{x}, \overline{y}\rangle \subset \mathcal{O}_{P,\mathcal{C}}$, whether P is simple or not. Moreover, since y is the tangent line to $\mathcal{C}$ at $P = (0,0)$ (recall that the tangent line at P is defined by the equation $\frac{\partial f}{\partial x}(P)x + \frac{\partial f}{\partial y}(P)y = 0$), we have

$$f = y + \text{higher terms}.$$

Grouping together terms involving y, we may write

$$f = yg - x^2h, \quad \text{where } g = 1 + \text{ higher terms, and } h \in K[x].$$

Thus, $\overline{yg} = \overline{x^2h} \in k[\mathcal{C}]$, so $\overline{y} = \overline{x^2h}\overline{g}^{-1}$ since $g(P) \neq 0$. It follows that $\mathcal{M}_{P,\mathcal{C}} = \langle \overline{x}, \overline{y} \rangle = \langle \overline{x} \rangle$, as desired.

Conversely, if $\mathcal{O} = \mathcal{O}_{P,\mathcal{C}}$ is a DVR, then $M = M_{P,\mathcal{C}} = \langle t \rangle$ for some prime $t \in M$. Thus, $\dim_K(M^n/M^{n+1}) = 1$ for all $n \geq 0$, and hence $m_P(\mathcal{C}) = 1$ by Theorem 4.5. This shows that P is simple.

Exercises

1. Complete the proof of Lemma 4.1.
2. Find all singular points in $\mathcal{C} = \mathbf{V}(x^2y^2 + x^2 + y^2 + 2xy(x+y+1)) \subset \mathbf{A}_{\mathbb{R}}^2$ and $\mathcal{C} = \mathbf{V}(xy + x^3 + y^3) \subset \mathbf{A}_{\mathbb{R}}^2$.
3. Show that the elliptic curve $y^2 = (x - a_1)(x - a_2)(x - a_3)$ in $\mathbf{A}_{\mathbb{C}}^2$, where a_1, a_2, a_3 are distinct complex numbers, has no singularities.
4. Use the notation as in the proof of Theorem 4.5 to show that the function $\chi(n) = \dim_K \left(\mathcal{O}_{P,\mathcal{C}}/M_{P,\mathcal{C}}^n \right)$ is a polynomial in n (for large n).

5. Normalization of Algebraic Curves

In the light of Proposition 3.6, Theorem 4.6 and Chapter 3 Theorem 4.7, we explore in this section the relation between the singularity of an algebraic curve $\mathcal{C}$ and the normality of the coordinate ring of $\mathcal{C}$. As a consequence, this leads to the "resolution" of singularities in algebraic curves by means of "normalization".

Let K be a field and let $\mathcal{C} = \mathbf{V}(f) \subset \mathbf{A}^2 = \mathbf{A}_K^2$ be an *irreducible plane curve*, where $f \in K[x,y]$ is a nonconstant polynomial. As before, $K[\mathcal{C}] = K[x,y]/\mathbf{I}(\mathcal{C})$ denotes the coordinate ring of $\mathcal{C}$ and $K(\mathcal{C})$ stands for the field of rational functions of $\mathcal{C}$ (i.e., the field of fractions of $K[\mathcal{C}]$). Put

$$\mathcal{S}(\mathcal{C}) = \left\{ P \in \mathcal{C} \mid P \text{ is singular} \right\}.$$

5.1. Theorem With notation as above, $\mathcal{S}(\mathcal{C})$ is a finite subset of $\mathcal{C}$. Hence $\mathcal{C} - \mathcal{S}(\mathcal{C})$, the set of nonsingular points in $\mathcal{C}$, is a dense open subset of $\mathcal{C}$.

Proof We may assume that f is irreducible and $\mathcal{S}(\mathcal{C}) \neq \emptyset$. Note that

$$\mathcal{S}(\mathcal{C}) = \mathbf{V}\left(f, \frac{\partial f}{\partial x}, \frac{\partial f}{\partial y}\right).$$

If $\frac{\partial f}{\partial x} \neq 0$, then since $\deg(\frac{\partial f}{\partial x}) < \deg f$ and f is irreducible, by Proposition 2.10, $\mathbf{V}\left(\frac{\partial f}{\partial x}, f\right)$ and hence $\mathcal{S}(\mathcal{C})$ is a finite set. A similar argumentation holds provided $\frac{\partial f}{\partial y} \neq 0$. Below we show that $\frac{\partial f}{\partial x}$ and $\frac{\partial f}{\partial y}$ cannot both be identically zero.

If $\mathrm{char}\,k = 0$, then $\frac{\partial f}{\partial x} = \frac{\partial f}{\partial y} = 0$ only when x and y do not appear in f. Hence f is a constant in K, a contradiction.

If $\mathrm{char}\,k = p > 0$, then $\frac{\partial f}{\partial x} = \frac{\partial f}{\partial y} = 0$ only when x and y appear in f in the form of $x^{\alpha p}$ and $y^{\beta p}$. Hence f can be expressed as $f = \sum a_{\alpha\beta} x^{p\alpha} y^{p\beta}$. Assuming that K is algebraically closed, the equation $a_{\alpha\beta} = t^p$ has a solution, say $a_{\alpha\beta} = s_{\alpha\beta}^p$ for some $s_{\alpha\beta} \in K$. Thus, putting $g = \sum s_{\alpha\beta} x^\alpha y^\beta$ we have $g^p = f$, contradicting the irreducibility of f. $\square$

In view of the last theorem, almost all points in $\mathcal{C}$ are simple. Next, we show that the normalization $\overline{K[\mathcal{C}]}$ of $K[\mathcal{C}]$ in $K(\mathcal{C})$ (i.e., the integral closure of $K[\mathcal{C}]$ in $K(\mathcal{C})$ in the sense of Chapter 3 (Definitions 1.6 and 3.1)) defines a nonsingular algebraic curve $\mathcal{C}'$ (see later Definition 5.5) which projects onto $\mathcal{C}$ with respect to the Zariski topology. To this end, we first generalize Chapter 4 Theorem 4.2(i) (concerning $\mathbb{Z} \subset \mathcal{A}_K$) to a more general setting.

5.2. Proposition Let $R \subseteq B$ be a module-finite ring extension, where R and B are domains. If every nonzero prime ideal of R is maximal, then every nonzero prime ideal of B is maximal.

Proof Suppose $B = \sum_{i=1}^m R\xi_i$ where $\xi_i \in B$. Let P be a nonzero prime ideal of B, $0 \neq b \in P$. Then b is integral over R, that is,

$$b^n + r_{n-1} b^{n-1} + \cdots + r_1 b + r_0, \quad r_i \in R.$$

Assume that n is the smallest degree. Then $0 \neq b_0 \in P \cap R$. Thus, $P \cap R$ is a nonzero prime ideal of R and hence maximal by the assumption. Put $K = R/(P \cap R)$. Then B/P may be viewed as an extension ring of K via the embedding $K = R/(P \cap R) \to B/P$. By Chapter 3 (section 3, exercise 6(b)), B/P is integral (indeed module-finite) over K. It follows from Chapter 3 Theorem 1.8 that B/P is a field. Therefore, P is maximal. $\square$

5.3. Corollary Let $C = \mathbf{V}(f) \subset \mathbf{A}^2$ be an irreducible plane curve, where $f \in K[x,y]$ is a nonconstant polynomial, and let $\overline{K[C]}$ be the normalization of $K[C]$ in $K(C)$. Then every nonzero prime ideal of $K[C]$, respectively every nonzero prime ideal of $\overline{K[C]}$, is maximal. In particular, $\overline{K[C]}$ is a Dedekind domain (in the sense of the final remark given in Chapter 4 section 4).

Proof By Corollary 2.12 we may assume that C is infinite. It follows from Proposition 1.8(i) and the Noether normalization (Chapter 3 Theorem 2.4) that $K[C]$ is module-finite over a polynomial subring $K[t]$. Furthermore by Chapter 3 Corollary 3.3, $\overline{K[C]}$ is module-finite over $K[C]$. Now Proposition 5.2 and Chapter 2 (section 1, exercise 2) yield the assertion. □

5.4. Theorem Let $C = \mathbf{V}(f) \subset \mathbf{A}^2$ be an irreducible plane curve, where $f \in K[x,y]$ is a nonconstant polynomial, and let $\overline{K[C]}$ be the normalization of $K[C]$ in $K(C)$. If K is algebraically closed, then the following statements are equivalent.
(i) C is nonsingular.
(ii) The local ring $\mathcal{O}_{P,C}$ of C at every $P \in C$ is a DVR.
(iii) $K[C] = \overline{K[C]}$, that is, $K[C]$ is a Dedekind domain.

Proof By Theorem 4.6, the equivalence (i) ⇔ (ii) holds even if K is not algebraically closed.

The implication (ii) ⇒ (iii) follows from Corollary 2.5(ii), Proposition 3.6, and Chapter 2 Theorem 3.1. And the implication (iii) ⇒ (ii) follows from Corollary 2.5(ii), Proposition 3.6, Chapter 2 Proposition 3.6, Chapter 3 (Theorem 4.1 and Corollary 4.6). □

Example (i) The fact that the cuspidal curve $C = \mathbf{V}(y^2 - x^3) \subset \mathbf{A}_\mathbb{R}^2$ has one singular point $(0,0)$ is indicated by $\mathbb{R}[C] \neq \overline{\mathbb{R}[C]}$ where the latter is isomorphic to the polynomial ring $\mathbb{R}[t]$ (see Chapter 3 (section 3, Example (iii))).

(ii) The fact that the nodal curve $C = \mathbf{V}(y^2 - x^3 - x^2) \subset \mathbf{A}_\mathbb{R}^2$ has one singular point $(0,0)$ is indicated by later section 7, Example (i) which shows $\mathbb{R}[C] \neq \overline{\mathbb{R}[C]} \cong \mathbb{R}[t]$.

Now let V be an algebraic set in the affine n-space $\mathbf{A}^n = \mathbf{A}_K^n$, $n \geq 1$. By the Noetherian normalization theorem, the coordinate ring $K[V]$ of V is module-finite over a polynomial subring $K[t_1,...,t_d]$, $d \leq n$. The last theorem leads to the following definition.

5.5. Definition Let $V \subset \mathbf{A}^n$ be an algebraic set, $n \geq 1$. If V is irreducible and its coordinate ring $K[V]$ is module-finite over a polynomial subring $K[t]$, then V is called an *algebraic curve* in $\mathbf{A}^n$.

Let $V \subset \mathbf{A}^n$ be an algebraic curve, $P \in V$. If the local ring $\mathcal{O}_{P,V}$ of V at P is a DVR then P is called a *nonsingular point* of V; otherwise, P is a *singular point* (or P defines a singularity in V). If every point of V is nonsingular then V is called a *nonsingular curve*.

Example (iii) By later exercise 1–2, the curve $\mathcal{C} = \mathbf{V}(y - f(x)) \subset \mathbf{A}^2_\mathbb{R}$ is nonsingular, where $f(x) \in K[x]$; and the twisted cubic curve $V = \mathbf{V}(y - x^2, z - x^3) \subset \mathbf{A}^3_\mathbb{R}$ is nonsingular.

5.6. Proposition Let $V \subset \mathbf{A}^n_K$ be an algebraic curve. If K is algebraically closed, then there exists a plane curve $\mathcal{C} \subset \mathbf{A}^2_K$ such that $K(V) = K(\mathcal{C})$. (Note that this qualifies Definition 5.5.)

Proof By the definition of an algebraic curve, $K[V] = \sum_{i=1}^m K[t]\xi_i$ where $K[t]$ is a polynomial subring of $K[V]$ and $\xi_i \in K[V]$. Then it is easy to see that $K(V) = K(t)[\xi_1, ..., \xi_m]$ where each ξ_i is algebraic over the field $K(t)$. We may assume that all ξ_i's are separable over $K(t)$. It follows from Chapter 1 Theorem 3.12 that $K(V) = K(t, \vartheta)$ for some $\vartheta \in K(V)$. Let $K[z_1, z_2]$ be the polynomial ring in z_1, z_2 over K and consider the onto ring homomorphism $K[z_1, z_2] \to K[t, \vartheta]$. Then $K[t, \vartheta] \cong K[z_1, z_2]/I$ for some ideal $I \subset K[z_1, z_2]$. Note that K is algebraically closed and hence infinite (Chapter 1 section 3, exercise 5). By the Nullstellensatz and Corollary 2.12, $K[z_1, z_2]/I$ is the coordinate ring of the plane curve $\mathcal{C} = \mathbf{V}(I)$. Hence $K(V) = K(\mathcal{C})$. □

5.7. Theorem (normalization of a plane curve) Let $\mathcal{C} = \mathbf{V}(f) \subset \mathbf{A}^2$ be an irreducible plane curve, where $f \in K[x, y]$ is a nonconstant polynomial, and let $\overline{K[\mathcal{C}]}$ be the normalization of $K[\mathcal{C}]$ in $K(\mathcal{C})$. If K is algebraically closed, then the following statements hold:

(i) There exists a nonsingular algebraic curve $V \subset \mathbf{A}^m$ for some $m \geq 1$ such that

$$K[V] = \overline{K[\mathcal{C}]}.$$

(ii) Let V be the algebraic curve obtained in part (i). There exists an onto polynomial mapping $\phi\colon V \to \mathcal{C}$ and an open subset $U \subset V$ such that $\phi(U) = \mathcal{C} - \mathcal{S}(\mathcal{C})$, the open subset of all nonsingular points in $\mathcal{C}$.

Proof (i) Assume that $\mathcal{C}$ is infinite. It follows from the proof of Corollary 5.3 that $\overline{K[\mathcal{C}]}$ is module-finite over a polynomial subring $K[t] \subset K[\mathcal{C}]$, say $\overline{K[\mathcal{C}]} = \sum_{i=1}^{s} K[t]\xi_i$, $\xi_i \in \overline{K[\mathcal{C}]}$. Hence $\overline{K[\mathcal{C}]} = K[t, \xi_1, ..., \xi_s] \cong K[z_1, ..., z_{s+1}]/I$ where I is an ideal of the polynomial ring $K[z_1, ..., z_{s+1}]$. Since K is algebraically closed, $K[z_1, ..., z_{s+1}]/I$ may be viewed as the coordinate ring of $V = \mathbf{V}(I) \subset \mathbf{A}_K^{s+1}$ by the Nullstellensatz. V is nonsingular because the localization of $\overline{K[\mathcal{C}]}$ at every nonzero prime ideal is a DVR by Corollary 5.3 and Chapter 3 Corollary 4.6.

(ii) Let α be the composite ring homomorphism

$$K[\mathcal{C}] \hookrightarrow \overline{K[\mathcal{C}]} \xrightarrow{\cong} K[V]$$

and suppose $\alpha(\overline{x}) = \overline{g}$, $\alpha(\overline{y}) = \overline{h} \in K[V]$, where $g, h \in K[z_1, ..., z_{s+1}]$. Thus, $\alpha(0) = \alpha(\overline{f}) = \overline{f(g,h)} = 0 \in K[V]$, that is, $f(g,h) \in I$. It follows that $0 = f(g,h)(Q) = f(g(Q), h(Q))$ for every $Q \in V$, and that there is a polynomial mapping

$$\phi : V \longrightarrow \mathcal{C}$$
$$Q \mapsto (g(Q), h(Q))$$

Now, if $P = (a, b) \in \mathcal{C}$ and $\langle x-a, y-b \rangle$ is the maximal ideal of P, then $\overline{g}-a$, $\overline{h}-b \in \langle \overline{z}_1 - c_1, ..., \overline{z}_{s+1} - c_{s+1} \rangle \subset K[V]$ by the weak Nullstellensatz and Chapter 3 Proposition 1.7, where $(c_1, ..., c_{s+1}) = Q \in V$. Thus, $g(Q) = a$, $h(Q) = b$ and hence $\phi(Q) = (a, b) = P$. This shows that ϕ is onto. Finally, since ϕ is continuous with respect to the Zariski topology by Proposition 2.7, Theorem 5.1 concludes that there is some open subset $U \subset V$ such that $\phi(U) = \mathcal{C} - \mathcal{S}(\mathcal{C})$. □

5.8. Definition The algebraic curve V obtained in Theorem 5.7 is called the *normalization* of $\mathcal{C}$, denoted $\overline{\mathcal{C}}$.

It follows from Chapter 3 Proposition 1.7 and part (ii) in the above proof of Theorem 5.7 that the following property holds for the normalization $\overline{\mathcal{C}}$ of an irreducible plane curve $\mathcal{C}$. This property will be a key to approach section 6 Theorem 6.3(ii).

5.9. Corollary With the assumption and notation as in Theorem 5.7, a maximal ideal of $K[\overline{\mathcal{C}}]$ contains no more than one maximal ideal of $K[\mathcal{C}]$.
□

Remark (i) With the discussion on algebraic curves given in this section, the interested reader can move forward to the *core* of algebraic curve theory — Riemann-Roch Theorem that deals with the divisor theory and consequently provides the way to compute one of the most important invariants – the *genus of* an algebraic curve.

(ii) For an irreducible algebraic set $V \subseteq \mathbf{A}^n$, after introducing the "dimension" of V and translating it into an algebraic version about $K[V]$ and $K(V)$, singularities in V can also be defined both geometrically and algebraically, and the singularity of a point $P \in V$ may be characterized by the property of $\mathcal{O}_{P,V}$ as well. In this lecture, of course, we do not go that far.

(iii) For an irreducible algebraic set $V \subseteq \mathbf{A}^n$, if we look at the normalization $\overline{K[V]}$ of $K[V]$ in $K(V)$, its structure is immediately clear by the Noether normalization, Chapter 3 (Corollary 3.3 and Theorem 4.7). The interested reader may start a study of the geometric impact of $\overline{K[V]}$ on V by means of exercise 3 below.

Exercises

1. Show that for the curve $\mathcal{C} = \mathbf{V}(y - f(x)) \subset \mathbf{A}_{\mathbb{R}}^2$, $\mathbb{R}[\mathcal{C}] \cong K[t]$, the polynomial ring in t over $\mathbb{R}$.
2. For the twisted cubic $V = \mathbf{V}(y - x^2, z - x^3) \subset \mathbf{A}_{\mathbb{R}}^3$, show that $\mathbb{R}[V] \cong K[t]$, the polynomial ring in t over $\mathbb{R}$.
   (Hint: In both exercises above apply section 2, exercise 11 to $t \mapsto (t, f(t))$, and $t \mapsto (t, t^2, t^3)$.)
3. Let V be an irreducible algebraic set in $\mathbf{A}_K^n$, and let $\overline{K[V]}$ be the normalization of $K[V]$. If K is algebraically closed, show that there is an irreducible algebraic set W such that $K[W] \cong \overline{K[V]}$ and W projects onto V with respect to the Zariski topology; moreover, every maximal ideal of $K[V]$ is contained in some maximal ideal of $\overline{K[V]}$, and every maximal ideal of $\overline{K[V]}$ contains no more than one maximal ideal of $K[V]$.

6. Parametrize a Rational Curve via Normalization

In section 2 we have remarked the significance of polynomial and rational parametrization. This section focuses on plane curves such that all but finitely many points can be rationally parametrized. Notations are maintained as before.

6.1. Definition Let $\mathcal{C} \subset \mathbf{A}^2 = \mathbf{A}_K^2$ be a plane curve, and let $K(t)$ be the field of fractions of the polynomial ring $K[t]$. $\mathcal{C}$ is called a *rational curve* if there is a rational parametrization (in the sense of Definition 2.6)

$$\phi: U \longrightarrow \mathcal{C}$$
$$c \mapsto \left(\frac{h_1(c)}{h_2(c)}, \frac{g_1(c)}{g_2(c)}\right) \quad U = \mathbf{A}_K^1 - (\mathbf{V}(h_2) \cup \mathbf{V}(g_2)), \ \frac{h_1}{h_2}, \frac{g_1}{g_2} \in K(t)$$

such that $\mathcal{C} - \phi(U)$ is finite.

Remark In the literature, a rational curve is usually defined by assuming that $\mathcal{C}$ is irreducible. Though Definition 6.1 above does not assume the irreducibility of $\mathcal{C}$, the next theorem tells that a rational curve defined in this way is necessarily irreducible.

6.2. Theorem Let $\mathcal{C} = \mathbf{V}(f) \subset \mathbf{A}^2$ be a rational plane curve with the rational parametrization ϕ as in Definition 6.1. If $\mathcal{C}$ is infinite then the following statements hold:
(i) $\phi(U)$ is open and dense in $\mathcal{C}$.
(ii) $\mathcal{C}$ is irreducible.
(iii) ϕ induces a K-linear ring isomorphism $K[\mathcal{C}] \cong K\left[\frac{h_1}{h_2}, \frac{g_1}{g_2}\right]$ and hence induces an injective K-linear ring homomorphism $K(\mathcal{C}) \to K(t)$.
(iv) $K(\mathcal{C}) \cong K(t)$.

Proof (i) The finiteness of $\mathcal{C} - \phi(U)$ implies $\phi(U)$ is open. Since every closed subset of $\mathbf{A}_K^1$ is finite and ϕ is continuous with respect to the Zariski topology (Proposition 2.7), we conclude that the closure $\overline{\phi(U)}$ of $\phi(U)$ in $\mathcal{C}$ is equal to $\mathcal{C}$, i.e., $\phi(U)$ is dense in $\mathcal{C}$.
(ii) By part (i), this follows from Proposition 2.7.
(iii) Consider the ring homomorphism

$$\psi: K[x,y] \longrightarrow K(t)$$
$$F \mapsto F\left(\frac{h_1}{h_2}, \frac{g_1}{g_2}\right)$$

which is clearly K-linear. We claim that $\text{Ker}\psi = \mathbf{I}(\mathcal{C})$. To see this, suppose $\psi(F) = F\left(\frac{h_1}{h_2}, \frac{g_1}{g_2}\right) = 0$. Then $F\left(\frac{h_1(c)}{h_2(c)}, \frac{g_1(c)}{g_2(c)}\right) = 0$ for all $c \in U$, and hence $F(\phi(U)) = 0$. Thus, $\phi(U) \subset \mathbf{V}(F) \cap \mathcal{C}$. But $\phi(U)$ is open and dense in $\mathcal{C}$ by part (i). Hence $F(\mathcal{C}) = 0$, i.e., $F \in \mathbf{I}(\mathcal{C})$. On the other hand, for

$G \in \mathbf{I}(\mathcal{C})$, $G(\phi(U)) = 0$ implies that $G\left(\frac{h_1(c)}{h_2(c)}, \frac{g_1(c)}{g_2(c)}\right) = 0$ for all $c \in U$, or equivalently, the rational function $r(t) = G\left(\frac{h_1}{h_2}, \frac{g_1}{g_2}\right) \in K(t)$ vanishes on U and hence on $\mathbf{A}_K^1 = K$. This shows that $0 = r(t) = \psi(G)$ because K is infinite (we assumed that $\mathcal{C}$ is infinite). Therefore, $\operatorname{Ker}\psi = \mathbf{I}(\mathcal{C})$, and $K[\mathcal{C}] \xrightarrow{\psi_*}_{\cong} K\left[\frac{h_1}{h_2}, \frac{g_1}{g_2}\right] \subset K(t)$ with $\psi_*(\overline{x}) = \frac{h_1}{h_2}$, $\psi(\overline{y}) = \frac{g_1}{g_2}$.

(iv) This follows from part (iii) and Lüroth's theorem (Chapter 1 Theorem 3.16). □

In view of section 4, if $\mathcal{C} \subset \mathbf{A}^2$ is a plane curve and $P \in \mathcal{C}$, then, after a change of coordinates we can always assume $P = (0,0) \in \mathcal{C}$. Thus, f is of the form

$$f = F_m + F_{m+1} + \cdots + F_q, \quad m \geq 1, \quad F_j = \sum_{\alpha+\beta=j} \lambda_{\alpha\beta} x^\alpha y^\beta,$$

where $m_P = m$ is the multiplicity of $\mathcal{C}$ at P.

6.3. Proposition Let K be an algebraically closed field, and let $\mathcal{C} = \mathbf{V}(f) \subset \mathbf{A}_K^2$ be a plane curve with $P = (0,0) \in \mathcal{C}$. Suppose $f = F_{m+1} + F_m$ with $m \geq 1$, as described above. If $F_m \neq \lambda x^m$ with $\lambda \in K^\times$ and $f(0,y)$ is not the zero polynomial, then $\mathcal{C}$ is a rational curve, and hence irreducible.

Proof Cutting the curve with the line $y = tx$ of variable slope t, or equivalently, substituting $y = tx$ into the equation $f(x,y) = 0$, we have

$$x^m F_m(1,t) + x^{m+1} F(1,t) = 0.$$

Then for $x \neq 0$,

$$(*) \qquad x = \frac{-F_m(1,t)}{F_{m+1}(1,t)}, \quad y = \frac{-tF_m(1,t)}{F_{m+1}(1,t)}.$$

Since K is algebraically closed and $F_m(1,t)$ is not a constant by the assumption, it follows that $P = (0,0)$ may be recaptured by $(*)$ above. Moreover, for any constant $c \in K$, $f(c,y)$ has only finitely many zeros. Hence $\mathcal{C}$ is a rational curve. □

Example (i) Clearly any curve of the form $\mathcal{C} = \mathbf{V}(y - f(x)) \subset \mathbf{A}_\mathbb{R}^2$ is rational, where $\phi \colon \mathbf{A}_\mathbb{R}^1 \to \mathcal{C}$ is defined by $\phi(c) = (c, f(c))$.

(ii) Every curve $\mathcal{C} \subset \mathbf{A}_\mathbb{C}^2$ defined by an irreducible quadratic polynomial $f \in \mathbb{C}[x, y]$ is rational.

(iii) Curves $\mathcal{C}_1 = \mathbf{V}(y^2 - x^3)$, $\mathcal{C}_2 = \mathbf{V}(y^2 - x^3 - x^2)$, $\mathcal{C}_3 = \mathbf{V}(y^3 - x^4 - x^3)$, $\mathcal{C}_4 = \mathbf{V}(x^3 + y^3 - 3xy)$ are rational in $\mathbf{A}_\mathbb{C}^2$.

Remark Not every irreducible plane curve is rational. For instance, after a change of coordinates, the polynomial $y^2 - x^3 - px - q$ with $4p^3 + 27q^2 \neq 0$ in $\mathbb{C}[x, y]$ becomes $y^2 - x(x - 1)(x - \lambda)$. By section 2, exercise 5, the curve $\mathcal{C} = \mathbf{V}(y^2 - x(x-1)(x-\lambda))$ is irreducible. But if $\text{char} K \neq 2$ and $\lambda \neq 0, 1$, then $\mathcal{C}$ is not rational. There are two different ways to know this.

One way may be explained as follows. If $\phi\colon U \to \mathcal{C}$ was a rational parametrization with $\phi(c) = \left(\frac{h_1(c)}{h_2(c)}, \frac{g_1(c)}{g_2(c)}\right)$ where $U = \mathbf{A}_K^1 - (\mathbf{V}(h_2) \cup \mathbf{V}(g_2))$, then by Theorem 6.2(iii) there would be an injective K-linear ring homomorphism $\phi_*\colon K(\mathcal{C}) \to K(t)$ with $\phi_*(\bar{x}) = \frac{h_1}{h_2}$, $\phi_*(\bar{y}) = \frac{g_1}{g_2}$, and thus there would be

$$\frac{g_1^2}{g_2^2} = \frac{h_1}{h_2}\left(\frac{h_1}{h_2} - 1\right)\left(\frac{h_1}{h_2} - \lambda\right), \quad \lambda \neq 0, 1.$$

Note that $h_1, h_2, g_1, g_2 \in K[t]$. One may further try to show that the above equality implies that $\frac{h_1}{h_2}, \frac{g_1}{g_2} \in K$.

Another way is to learn a theory on the genus of curves. The theory on genus asserts that an irreducible plane curve $\mathcal{C}$ is rational if and only if its genus $g(\mathcal{C})$ is zero, where if $\mathcal{C}$ is defined by a polynomial of degree m, then

$$g(\mathcal{C}) = \frac{(m-1)(m-2)}{2} - \#(\text{singularities of } \mathcal{C} \text{ properly counted}).$$

By Theorem 6.2, if $\mathcal{C}$ is a rational curve, then $K(\mathcal{C}) \cong K(t)$. We now consider the converse by passing to the normalization of $\mathcal{C}$. Part (ii) of the next theorem seems missing in the literature.

6.4. Theorem Let $\mathcal{C} = \mathbf{V}(f) \subset \mathbf{A}_K^2$ be an irreducible plane curve, and suppose that $\mathcal{C}$ is infinite.
(i) If there is an injective K-linear ring homomorphism $K(\mathcal{C}) \to K(t)$, then there is a rational parametrization for $\mathcal{C}$.
(ii) (Huishi Li) If K is algebraically closed and $K(\mathcal{C}) \cong K(t)$, then there exists a rational parametrization ϕ as in Definition 6.1 such that either $\mathcal{C} - \phi(U) = \emptyset$ or $\mathcal{C} - \phi(U)$ contains only one point. Therefore $\mathcal{C}$ is rational.

Proof (i) By the assumption we have the following composite ring homomorphism

$$\varphi: \quad K[\mathcal{C}] \hookrightarrow K(\mathcal{C}) \longrightarrow K(t)$$

with $\varphi(\overline{x}) = \frac{h_1}{h_2}$, $\varphi(\overline{y}) = \frac{g_1}{g_2} \in K(t)$. Set $U = K - \mathbf{V}(h_2) \cup \mathbf{V}(g_2)$. Then since $0 = \varphi(f) = f\left(\frac{h_1}{h_2}, \frac{g_1}{g_2}\right)$, the latter defines a rational parametrization:

$$\psi: \quad U \longrightarrow \quad \mathcal{C}$$

$$c \mapsto \left(\frac{h_1(c)}{h_2(c)}, \frac{g_1(c)}{g_2(c)}\right)$$

(ii) Let $K[\overline{\mathcal{C}}]$ be the coordinate ring of the normalization $\overline{\mathcal{C}}$ of $\mathcal{C}$, and consider the composite ring homomorphism

$$\varphi: \quad K[\mathcal{C}] \hookrightarrow K[\overline{\mathcal{C}}] \hookrightarrow K(\mathcal{C}) \xrightarrow[\cong]{\alpha} K(t)$$

Let $\psi: U \to \mathcal{C}$ be the parametrization as obtained in part (i). If $P = (a, b) \in \mathcal{C}$, and $\langle \overline{x}-a, \overline{y}-b \rangle$ is the maximal ideal of P in $K[\mathcal{C}]$, then $\langle \overline{x}-a, \overline{y}-b \rangle \subset M$ for some maximal ideal M of $K[\overline{\mathcal{C}}]$ by Chapter 3 Proposition 1.7. Let R be the DVR in $K(t)$ that corresponds to $K[\overline{\mathcal{C}}]_M$, where the latter is the localization of $K[\overline{\mathcal{C}}]$ at M, which, by Corollary 5.3 and the proof of Chapter 2 Theorem 2.10, is a DVR containing K and has its field of fractions $K(\mathcal{C})$. By Chapter 2 (section 2, Example (vii)), the only DVRs in $K(t)$ that contain K and have the field of fractions $K(t)$ are those localizations $K[t^{-1}]_{\langle t^{-1} \rangle}$ and $K[t]_{\langle t-\lambda \rangle}$, $\lambda \in K$. So, if $\alpha(K[\overline{\mathcal{C}}]_M) = R \neq K[t^{-1}]_{\langle t^{-1} \rangle}$ then $\alpha(K[\overline{\mathcal{C}}]_M) = R = K[t]_{\langle t-\lambda \rangle}$ for some $\lambda \in K$ (weak Nullstellensatz). In the second case we have $\frac{h_1}{h_2} - a$, $\frac{g_1}{g_2} - b \in \langle t-\lambda \rangle \subset K[t]_{\langle t-\lambda \rangle}$. Hence $\frac{h_1}{h_2}$, $\frac{g_1}{g_2}$ are defined at λ and

$$\psi(\lambda) = \left(\frac{h_1(\lambda)}{h_2(\lambda)}, \frac{g_1(\lambda)}{g_2(\lambda)}\right) = (a, b) = P.$$

By Corollary 5.9, at most one maximal ideal of $K[\mathcal{C}]$ is contained in some maximal ideal M of $K[\overline{\mathcal{C}}]$ where $K[\overline{\mathcal{C}}]_M$ corresponds to $K[t^{-1}]_{\langle t^{-1} \rangle}$. Therefore, there is at most one $P = (a, b) \in \mathcal{C}$ which is not an image of the rational mapping ψ defined in part (i).

Exercises

1. Find a rational parametrization of the curve $C = \mathbf{V}(y^2 - x^4 - x^2) \subset \mathbf{A}_\mathbb{R}^2$.
   (Hint: Use the line $y = tx$ to cut C first and then use the line $t = u(x-1)$ to cut the obtained hyperbola $C' = \mathbf{V}(x^2 - t^2 - 1)$.)
2. Show that the curve $C = \mathbf{V}(y^2(a-x) - x^3) \subset \mathbf{A}_\mathbb{R}^2$ is rational, where $a \in \mathbb{R}$.
3. Show that the curve $C = \mathbf{V}((x^2 + y^2)^2 + 3x^2 y - y^3) \subset \mathbf{A}_\mathbb{C}^2$ is rational.

7. Rational Curves and Diophantine Equations

In number theory one of the themes is to find the integer solutions of the Diophantine equation $f(x_1, ..., x_n) = 0$ where $f \in \mathbb{Q}[x_1, ..., x_n]$, or in the language of algebraic geometry, is to find the points $P = (a_1, ..., a_n)$ on the hypersurface $\mathbf{V}(f) \subset \mathbf{A}_\mathbb{R}^n$ with all $a_i \in \mathbb{Z}$. The subject of rational curves leads in a natural way to some connection with this aspect, though not every curve is rational. We close this course by examples that illustrate such a point of view.

Example (i) Consider the nodal curve $C = \mathbf{V}(y^2 - x^3 - x^2) \subset \mathbf{A}_\mathbb{R}^2$.

After cutting the curve with the line $y = tx$ of variable slope t, we obtain $t^2 x^2 = x^3 + x^2$. It follows that if $x \neq 0$ then

(1) $$\begin{cases} x = t^2 - 1 \\ y = t(t^2 - 1) \end{cases}$$

But $t = \pm 1$ also gives $(0, 0)$ in C. This shows that every point of C can be obtained by system (1) with $t \in \mathbb{R}$. By Definition 6.1 and Theorem 6.2, C is rational and irreducible. Moreover, putting $h = t^2 - 1$, $g = t(t^2 - 1)$, then $\mathbb{R}[C] \cong \mathbb{R}[h, g] \subset \mathbb{R}(t)$ by Theorem 6.2. Thus, it is clear that $\mathbb{R}[C] \neq \overline{\mathbb{R}[C]} \cong \mathbb{R}[t]$.

If $P = (a, b) \in C$ with $a, b \in \mathbb{Z}$, then we say that P is an integral point. By system (1) it is clear that C has infinitely many integral points if we let $t \in \mathbb{Z}$. Conversely, suppose (a, b) is an integral point. We may assume $a \neq 0$ (note that $(0, 0)$ can be obtained from system (1)). Then $t = \frac{b}{a}$ is a rational number but $t^2 = a + 1$ is an integer by system (1). Thus, t must be an integer, and hence all integral points on C are given by system (1) with $t \in \mathbb{Z}$. In other words, we obtain all integer solutions of the Diophantine equation $y^2 - x^3 - x^2 = 0$ by letting $t \in \mathbb{Z}$ in the equation system (1).

(ii) Similarly as dealing with Example (i), the curve $C = \mathbf{V}(x^6 - x^2y^3 - y^5) \subset \mathbf{A}_{\mathbb{R}}^2$ may be parametrized by intersecting it with the line $y = tx$ of variable slope t:

(2)
$$\begin{cases} x = t^3 + t^5 \\ y = t^4 + t^6 \end{cases}$$

Every point of C is given by system (2) with $t \in \mathbb{R}$, C is irreducible and rational.

Clearly, if $t \in \mathbb{Z}$ then from system (2) we get integral points on the curve. We claim that these are all integral points on the curve. Suppose (x, y) is an integral point. Then x and $y = tx$ are integers. Since $x = t^3(1+t^2)$, the only way that x can be zero is for t to be zero. Thus it may be assumed that $x \neq 0$ and so $t = \frac{y}{x}$ is rational. We write t in the form $t = \frac{a}{b}$ where a and b are relatively prime integers. Now $x = t^3 + t^5$ and so $x = \frac{a^3b^2 + a^5}{b^5}$. This must be an integer and so $b|(a^3b^2 + a^5)$. Therefore $b|a^5$. Since a and b are relatively prime, it follows that $b = \pm 1$. Thus t must be an integer, as we claimed. Therefore, all integral points in C can be obtained by system (2) with $t \in \mathbb{Z}$. In other words, we obtain all integer solutions of the Diophantine equation $x^6 - x^2y^3 - y^5 = 0$ by letting $t \in \mathbb{Z}$ in the system (2).

(iii) It is well known that the unit circle $C = \mathbf{V}(x^2 + y^2 - 1) \subset \mathbf{A}_{\mathbb{R}}^2$ can be parametrized by using trigonometric functions:

$$\begin{cases} x = cos(t), \\ y = sin(t). \end{cases}$$

However, if we intersect the circle with the line $y = tx - 1$ of variable slope t (using $y = tx$ will cut the circle in two points), we obtain the rational parametrization of C:

(3)
$$\begin{cases} x = \dfrac{2t}{1+t^2}, \\ y = \dfrac{1-t^2}{1+t^2}. \end{cases}$$

Note that the system (3) does not describe the whole circle since $y = \frac{1-t^2}{1+t^2}$ can never equal -1, that is, the point $(0, -1)$ is not covered. However, by Definition 6.1 and Theorem 6.2 we know that C is irreducible and rational. We now use the system (3) to find the integer solutions (X, Y, Z) of the Diophatine equation $Z^2 + Y^2 = Z^2$ with $\gcd(X, Y, Z) = 1$ (recall the argumentation on this problem given in the introduction part of Chapter 4), that is the same as to find the rational solutions $\left(x = \frac{X}{Z}, y = \frac{Y}{Z}\right)$ for the equation $x^2 + y^2 = 1$. From (3) it is clear that x and y are rational if and only if t is rational. So assume that $t = \frac{u}{v}$ where $u, v \in \mathbb{Z}$ and $\gcd(u, v) = 1$, and for the moment assume that X, Y, and Z are positive. Thus,

$$x = \frac{X}{Z} = \frac{v^2 - u^2}{v^2 + u^2}, \quad y = \frac{Y}{Z} = \frac{2uv}{v^2 + u^2}$$

and since $\gcd(X, Y) = \gcd(X, Z) = \gcd(Y, Z) = 1$, there is a positive integer m such that

$$mZ = v^2 + u^2, \quad mX = v^2 - u^2, \quad mY = 2uv.$$

We claim that $m = 1$. To see this, note that $m|(v^2 + u^2)$, $m|(v^2 - u^2)$. It follows that $m|(2v^2)$, $m|(2u^2)$. But $\gcd(u, v) = 1$. So $m = 1$ or $m = 2$. Since X and Y cannot both be odd or even (otherwise $4|(Z^2 - 2)$), we may assume Y is even. Then, $2|(uv)$ and one of u, v is even. Consequently, one of u^2, v^2 is even and the other is odd. Hence $u^2 + v^2$ is odd, contradicting $2Z = u^2 + v^2$. Therefore, $m = 1$. This shows that all desired integer solutions (X, Y, Z) are given by

$$X = \pm(v^2 - u^2), \quad Y = \pm 2uv, \quad Z = \pm(v^2 + u^2).$$

Exercises

1. Indicate why in Example (iii) X and Y cannot both be odd or even.
2. Find the integer solutions of the equation $y^3 - x^4 - x^3 = 0$.

References

[Coh] P. M. Cohn, *Algebra* I & II, John Wiley and Sons Ltd., 1982.
[Edw] H. M. Edwards, *Fermat's Last Theorem*, Springer, 1997.
[Ful] W. Fulton, *Algebraic Curves*, W. A. Benjamin, New York, 1969.
[Jac] N. Jacobson, *Basic Algebra*, Vol. II, San Francisco, 1974–1980.
[LVO] H. Li and F. Van Oystaeyen, *A Primer of Algebraic Geometry*, Marcel Dekker, 2000.
[Mar] D. A. Marcus, *Number Fields*, Springer-Verlag, 1977.
[Mat] H. Matsumura, *Commutative Algebra* (second edition), The Benjamin/Cummings Publishing Company, Inc., 1980.
[Rei] M. Reid, *Undergraduate Commutative Algebra*, LMS, Student Texts 29, 1995.
[Sam] P. Samuel, *Algebraic Theory of Numbers*, English translation (by A. J. Silberger), Hermann, 1971.
[ST] I. N. Stewart and D. O. Tall, *Algebraic Number Theory* (second edition), Chapman and Hall, 1987.
[Van] B. L. Van der Waerden, *Algebra* I & II, Springer-Verlag, 1985.

Index

affine space, 127
algebra, 79
algebra homomorphism, 79
algebraic closure, 28
algebraic curve, 160
algebraic element, 22
algebraic field extension, 22
algebraic integer, 102
algebraic set, 128
algebraically closed, 25
algebraically independent, 86
ascending chain condition, 6
associate, 9
associated prime, 95

basis, 40, 48
bilinear form, 37

characteristic, 3
closed set, 144
closure of a subset, 145
continuous function, 145
contraction of an ideal, 71
coordinate ring, 136
cuspidal curve, 128
cyclotomic field, 109

Dedekind domain, 125
degree of a curve, 151
degree of a field extension, 22
degree of a polynomial, 4
dense subset, 145

direct sum of modules, 48
discrete valuation, 60
discrete valuation ring, 61
discriminant, 104
divisible, 9
divisor, 9
domain, 3
DVR, 61

extension field, 18
extension of an ideal, 71

field, 3
field extension, 18
field of fractions, 4
field of rational functions, 148
finitely generated extension ring, 80
finitely generated ideal, 2
finitely generated module, 47
finitely generated subring, 2
fractional ideal, 119
free abelian group, 40
free module, 48

group of units, 9

hypersurface, 128

ideal of an algebraic set, 129
induced topology, 145
inseparable field extension, 24
inseparable polynomial, 22

integral basis, 102
integral closure, 83
integral element, 80
integral extension, 80
integrally closed, 83
irreducible, 9
irreducible algebraic set, 135
irreducible component, 142

Jacobson radical, 57

local ring, 58
local ring at P, 149
localization at P, 72

maximal condition, 6
maximal ideal, 53
maximal spectrum, 53
minimal polynomial, 23
module, 45
module homomorphism, 47
module of fractions, 75
module-finite, 80
monic polynomial, 5
multiplicative set, 53
multiplicity at P, 155

neighborhood of a point, 144
nilradical, 56
nodal curve, 128
Noetherian module, 49
Noetherian ring, 6
Noetherian space, 141
nonsingular curve, 153
nonsingular point, 153, 160
norm, 35
normal domain, 90
normalization, 90
normalization of a curve, 161
number field, 102

open set, 144

p-adic valuation, 62
PID, 7
plane curve, 128

pole set, 147
polynomial function, 146
polynomial mapping, 138
polynomial parametrization, 139
prime, 11
prime ideal, 53
prime spectrum, 53
prime subfield, 3
primitive element, 25
primitive polynomial, 14
principal ideal ring, 7
product topology, 145
proper factorization, 9

quadratic number field, 108
quotient module, 47

R-basis, 48
radical ideal, 132
rational curve, 163
rational mapping, 139
rational parametrization, 139
reducible, 9
ring extension, 79
ring of algebraic integers, 102
ring of fractions, 68

separable field extension, 24
separable polynomial, 22
simple extension field, 18
simple point, 153
singular point, 153
splitting field, 19
square-free, 80
subalgebra, 79
submodule, 46
symmetric polynomial, 29

tangent line, 153
topological space, 144
topology, 144
total polynomial, 34
trace, 35
transcendental element, 22
transcendental field extension, 22
trivial divisor, 9

twisted cubic, 128

UFD, 11
unimodular, 41
unit, 3

valuation ring, 65

Z-basis, 40
Zariski topology, 134